Lecture Notes in Statistics

135

Edited by P. Bickel, P. Diggle, S. Fienberg, K. Krickeberg,
I. Olkin, N. Wermuth, S. Zeger

Springer

New York
Berlin
Heidelberg
Barcelona
Budapest
Hong Kong
London
Milan
Paris
Singapore
Tokyo

Christian P. Robert (Editor)

Discretization and MCMC Convergence Assessment

 Springer

Christian P. Robert
INSEE
Crest
Batiment Malakoff 1-Timbre J037
92245 Malakoff Cedex
France

Library of Congress Cataloging-in-Publication Data
Discretization and MCMC convergence assessment / Christian P. Robert
 (editor).
      p.   cm. -- (Lecture notes in statistics ; 135)
   Includes bibliographical references and indexes.
   ISBN 0-387-98591-3 (alk. paper)
   1. Convergence. 2. Markov processes. 3. Monte Carlo method.
I. Robert, Christian P., 1961-   . II. Series: Lecture notes in
statistics (Springer-Verlag) ; v. 135.
QA295.D46   1998
519.2--dc21   98-8676

Printed on acid-free paper.

Camera ready copy provided by the editor.
Printed and bound by Braun-Brumfield, Ann Arbor, MI.
Printed in the United States of America.

9 8 7 6 5 4 3 2 1

ISBN 0-387-98591-3 Springer-Verlag New York Berlin Heidelberg   SPIN 10687155

Preface

The exponential increase in the use of MCMC methods and the corresponding applications in domains of even higher complexity have caused a growing concern about the available convergence assessment methods and the realization that some of these methods were not reliable enough for all-purpose analyses. Some researchers have mainly focussed on the convergence to stationarity and the estimation of rates of convergence, in relation with the eigenvalues of the transition kernel. This monograph adopts a different perspective by developing (supposedly) practical devices to assess the mixing behaviour of the chain under study and, more particularly, it proposes methods based on finite (state space) Markov chains which are obtained either through a discretization of the original Markov chain or through a duality principle relating a continuous state space Markov chain to another finite Markov chain, as in missing data or latent variable models.

The motivation for the choice of finite state spaces is that, although the resulting control is *cruder*, in the sense that it can often monitor convergence for the discretized version alone, it is also much *stricter* than alternative methods, since the tools available for finite Markov chains are universal and the resulting transition matrix can be estimated more accurately. Moreover, while some setups impose a fixed finite state space, other allow for possible refinements in the discretization level and for consecutive improvements in the convergence monitoring.

An exciting, though challenging, approach to both Markovian discretization and convergence monitoring is based on *renewal theory*. The monograph introduces the useful features of this theory (small sets, minorization conditions, renewal time, regeneration, etc.) and focuses on the discretization aspect, which is to consider in which small set the (continuous) Markov chain is at each renewal time. We also mention some general control device which is based on the stability of the limiting variance in the CLT.

Once a finite Markov chain is constructed, there are many convergence results and assessment controls available, and this monograph selects those which are the most well-grounded and potentially implementable. Two approaches are for instance borrowed from Kemeny and Snell (1960), based on divergence and asymptotic variance, both of which relate to the transition matrix of the chain and a transformation of this matrix called the fundamental matrix. The derived convergence assessments are then to monitor the stabilizing of the empirical divergences and of some asymptotic variance evaluations, plus to wait for the agreement with the corresponding theoretical limits, evaluated from the estimated transition matrix.

A side-effect of this study is to expose the importance of *coupling* on both convergence assessment and convergence control. In fact, coupling strategies are usually easier to implement for finite Markov chains. We show how coupling can accelerate convergence, in particular in the case parallel chains are used to evaluate different divergences. We also examine some issues related with perfect sampling, although this is not the topic of the book.

The Central Limit Theorem, which holds in most finite setups, can be used as a convergence assessment tool, in the sense that normality of normalized sums can be tested on parallel chains. We also take advantage of subsampling techniques to use the standard CLT.

The different methods proposed in the monograph are evaluated on a set of *benchmark* examples in order to keep the comparison going. These examples involve the aggregated multinomial model of Tanner and Wong (1987), the nuclear pump failure model of Gaver and O'Muircheartaigh (1987) and a Cauchy model of Robert (1995). In order to extend this evaluation to more realistic setups, we also propose in the last chapters of the monograph full scale applications to analysis and prediction for DNA sequences, which attacks the identification of homogeneous regions in DNA sequences by a hidden Markov chain modelling and involves high dimension Markov chains, latent structures for the dynamics of HIV infection with measurement errors and a hierarchical Bayes structure, as well as the estimation of exponential mixtures applied to a hospitalization dataset.

The monograph is the outcome of a monthly research seminar held at CREST (Paris) since 1995, which involved the authors of the different chapters, as well as Gilles Celeux, Jérôme Dupuis and Marc Lavielle, to which we are indebted for comments and criticisms. Partial financial support from CREST must also be acknowledged. At last, Costas Goutis came and participated to one of our seminars. Since he has left us to hike in other galaxies on July 22, 1996, I would like to dedicate this monograph to his memory.

Paris, April 1998

Christian P. Robert

Contents

List of Contributors

Dominique Cellier
Laboratoire Analyse et Modèles
Stochastiques – UPRESA 6085
Université de Rouen
76821 Mont Saint Aignan cedex
France
cellier@ams.univ-rouen.fr

Didier Chauveau
Equipe d'Analyse et de Mathématiques
Appliquées
Université de Marne-la-Vallée
Citée Descartes
5, boulevard Descartes
77454 Marne-la-Vallée cedex 2
France
chauveau@math.univ-mlv.fr

Jean Diebolt
CNRS et Imag/LMC/Statistique
et Modélisation Stochastique
Université de Grenoble
BP 53
38041 Grenoble cedex 9, France
jed@ccr.jussieu.fr

Marie-Anne Gruet
Département de Biométrie
et Intelligence Artificielle
INRA – Domaine de Vilvert
78352 Jouy-en-Josas cedex, France
mag@baobab.jouy.inra.fr

Chantal Guihenneuc-Jouyaux
Laboratoire de Statistique Médicale
– URA CNRS 1323
Université Paris V
45 rue des Saints-Pères
75006 Paris, France
guihenne@citi2.fr

Virginie Lasserre
Laboratoire de Statistique Médicale
– URA CNRS 1323
Université Paris V
45 rue des Saints-Pères
75006 Paris, France
lasserre@citi2.fr

Florence Muri
Laboratoire de Statistique Médicale
– URA CNRS 1323
Université Paris V
45 rue des Saints-Pères
75006 Paris, France
muri@citi2.fr

Anne Philippe
Laboratoire de Probabilités
et Statistique, CNRS E.P. 1765
UFR de Mathématiques – Bât. M2
Université Lille I
59655 Villeneuve d'Ascq cedex
France
Anne.Philippe@univ-lille1.fr

Sylvia Richardson
Unité 170, INSERM
16, avenue Paul Vaillant Couturier
94807 Villejuif cedex
France
richardson@vjf.inserm.fr

Christian P. Robert
Laboratoire de Statistique
CREST – INSEE
75675 Paris cedex 14, France
robert@ensae.fr

1
Markov Chain Monte Carlo Methods

Christian P. Robert
Sylvia Richardson

1.1 Motivations

As the complexity of the models covered by statistical inference increases, the need for new computational tools gets increasingly pressing. Simulation has always been a natural tool for statisticians (as opposed to numerical analysis) and simulation via Markov chains has been recently exposed as a broad spectrum method, which allows to tackle problems of higher complexity (as shown by the subsequent literature). Although the purpose of this book is to introduce some control techniques for such simulation methods, we feel it is necessary to recall in this chapter the main properties of Markov Chain Monte Carlo (MCMC) algorithms. Moreover, we take the opportunity to introduce notations and our favorite (so-called *benchmark*) examples, which will be used over and over in the first half of the book. Good introductions to the topic are Gelfand and Smith (1990) seminal paper and Tanner (1996) monograph, as well as Casella and George (1992) and Chib and Greenberg (1996) tutorial papers, and Gelman and Rubin (1992), Geyer (1992) and Besag *et al.* (1995) surveys, while Neal (1993), Gilks, Richardson and Spiegelhalter (1996), Robert (1996c), Gamerman (1997), Robert and Casella (1998) and Gelfand and Smith (1998) provide deeper entries. In this explosive area of research, many books, monographs and long surveys are currently on their way and it is quite impossible to keep an exact account of the current[1] MCMC production!

1.2 Metropolis-Hastings algorithms

Consider a distribution with density function f, which is complex enough for integrals of the form

$$\int h(x)f(x)dx \tag{1.1}$$

---

[1] As an illustration, visit the site `http://www.stats.bris.ac.uk/MCMC/pages` for a sample of the latest publications on the topic.

to be unavailable in closed form. Assume in addition that f is such that regular Monte Carlo evaluation cannot apply, in the sense that a random generator from f is not readily available. The density f is also unavailable in closed form and prevents from using an importance sampling method. The fundamental idea behind the MCMC algorithms is then to approximate (1.1) via an ergodic Markov chain $(x^{(t)})$ with stationary distribution f. For T_0 "large enough", $x^{(T_0)}$ is roughly distributed from f and the sample $x^{(T_0)}, x^{(T_0+1)}, \ldots$ can be used for most purposes as an iid sample from f, even though the $x^{(T_0+t)}$'s are not independent. For instance, the Ergodic Theorem (Revuz, 1984, or Meyn and Tweedie, 1993) justifies the approximation of (1.1) by the empirical average

$$\frac{1}{T}\sum_{t=1}^{T} h(x^{(T_0+t)}) \, , \tag{1.2}$$

in the sense that (1.2) is converging to (1.1) for almost every realization of the chain $(x^{(t)})$ under minimal conditions. The goal of this book is to provide general techniques to assess convergence of either the average (1.2) to the integral (1.1) or of the chain $(x^{(t)})$ to its stationary distribution, borrowing from finite Markov chain theory. We thus refrain from discussing the theoretical aspects of MCMC algorithms when they are not relevant for control purposes, referring the reader to the huge literature on the topic (see, e.g., Athreya *et al.*, 1996, Gelfand and Smith, 1998, Mengersen and Tweedie, 1996, Roberts and Rosenthal, 1997, Roberts and Tweedie, 1996, Tierney, 1994).

At this point, the use of a Markov chain with stationary distribution f may sound quite formal or even artificial as f cannot be directly simulated and the construction of this chain has yet to be explained. We now present a very general class of transition kernels with stationary distribution f, based on the *Metropolis–Hastings scheme*. This method has been developed by Metropolis, Rosenbluth, Rosenbluth, Teller and Teller (1953) in an optimization setup with a discrete state space and applied in a statistical setting by Hastings (1970) and Peskun (1973). Although others (see, e.g., Geman and Geman, 1984; Tanner and Wong, 1987; Besag, 1989) have also stressed the relevance of a simulation approach based on Markov chains, the introduction of Gibbs sampling techniques by Gelfand and Smith (1990) is widely acknowledged as the actual starting point of these methods in Statistics, particularly in Bayesian Statistics.

The Metropolis–Hastings algorithm is based on a general conditional density $q(y|x)$, called *instrumental distribution*. The choice of $q(\cdot|x)$ is almost unrestricted but, obviously, a practical requirement is that simulations from $q(\cdot|x)$ should be easily obtainable. In addition, it is necessary that either $q(\cdot|x)$ is available in closed form (up to a multiplicative constant independent from x), or is symmetric, i.e. such that $q(x|y) = q(y|x)$. The algorithm then proceeds to construct a Markov chain $(x^{(t)})$ as follows: given $x^{(t)}$,

1. **Generate** $y_t \sim q(y|x^{(t)})$.

2. **Take**

$$x^{(t+1)} = \begin{cases} y_t & \text{with probability} \quad \rho(x^{(t)}, y_t), \\ x^{(t)} & \text{with probability} \quad 1 - \rho(x^{(t)}, y_t), \end{cases}$$

$$[A_1]$$

where

$$\rho(x^{(t)}, y_t) = \min\left\{ \frac{f(y_t)}{f(x^{(t)})} \frac{q(x^{(t)}|y_t)}{q(y_t|x^{(t)})}, 1 \right\}.$$

Other choices for $\rho(x^{(t)}, y_t)$ are possible (see Hastings, 1970, Peskun, 1973, or Neal, 1993), but the one above enjoys some (limited) optimality (see Mira and Geyer, 1998) and is usually preferred in practice. The algorithm thus always accepts simulations y_t with a ratio $f(y_t)/q(y_t|x^{(t)})$ larger than $f(x^{(t)})/q(x^{(t)}|y_t)$. Note also that $\rho(x^{(t)}, y_t)$ only depends on $f(y_t)/f(x^{(t)})$ and $q(x^{(t)}|y_t)/q(y_t|x^{(t)})$ and is thus independent of normalization constants. At last, if f is already a stationary distribution for the transition kernel, i.e. if

$$f(x^{(t)})q(y_t|x^{(t)}) = f(y_t)q(x^{(t)}|y_t),$$

the acceptance probability is always equal to 1, which shows that the Metropolis–Hastings algorithm is making optimal use of the information contained in the couple $(f, q(\cdot|x))$.

The validity of $[A_1]$, i.e. the fact that f is truly a stationary distribution for the chain $(x^{(t)})$, follows from the so-called *detailed balance condition*

$$f(x^{(t)})q(y_t|x^{(t)}) = f(y_t)q(x^{(t)}|y_t),$$

which is indeed satisfied (and, furthermore, ensures the *reversibility* of the chain) and from the simple additional condition on $q(\cdot|x)$

$$\text{supp}(f) \subset \text{supp}\{q(\cdot|x)\} \tag{1.3}$$

for a.e. $x \in \text{supp}(f)$. Obviously, if (1.3) does not hold, it is impossible to explore the whole support of f and the stationary distribution is then the restriction of f to some subset. The ergodicity of $(x^{(t)})$ indeed follows from its *f-irreducibility* and its *aperiodicity*, as shown by Tierney (1994). We refer to Athreya, Doss and Sethuraman (1996) and Roberts and Tweedie (1996) for more advanced theoretical studies of the probabilistic properties of Metropolis–Hastings chains.

Again, a fundamental feature about Metropolis–Hastings algorithms is that they can use any conditional distribution $q(\cdot|x)$ provided the support condition (1.3) holds. Two particular types of distributions are often singled out, leading to the independent and the random walk Metropolis–Hastings algorithms. The *independent Metropolis–Hastings algorithm* is based upon an instrumental distribution q which does not depend on $x^{(t)}$. Denoting by g this distribution, that is $q(y|x^{(t)}) = g(y)$, the acceptance probability in $[A_1]$ is then

$$\varrho(x^{(t)}, y_t) = \min\left\{\frac{f(y_t)\, g(x^{(t)})}{f(x^{(t)})\, g(y_t)}, 1\right\}.$$

Mengersen and Tweedie (1996) provide a detailed analysis of this algorithm, showing in particular that geometric ergodicity requires the ratio f/g to be bounded a.e., as in accept-reject algorithms (see Ripley, 1987, or Robert, 1996c). (See Liu, 1995, and Casella and Robert, 1996, for a comparison between both algorithms.) The *random walk Metropolis–Hastings algorithm* is based on a "local" perturbation of the previous value of $x^{(t)}$, i.e. $y_t = x^{(t)} + \epsilon_t$, thus with $q(y|x)$ of the form $g(y - x)$. An interesting property follows from the choice of a symmetric function g (i.e. such that $g(-t) = g(t)$) since the acceptance probability only depends on the ratio $f(y_t)/f(x^{(t)})$, as in Metropolis *et al.* (1953).

Example 1.2.1 In his seminal paper, Hastings (1970) considers generating the Gaussian distribution $\mathcal{N}(0, 1)$ from the uniform distribution on $[-\delta, \delta]$. The acceptance probability is then $\rho(x_t, y_t) = \exp\{(x_t^2 - y_t^2)/2\} \wedge 1$. Note that values of y_t smaller than $x^{(t)}$ are always accepted, while the choice of δ dictates the speed of exploration of the support of f.   ‖

Surprisingly, despite its intuitive appeal (and also good performances in practice), the random walk Metropolis–Hastings algorithm may have poor convergence properties. Mengersen and Tweedie (1996) have indeed shown that, when supp $f = \mathbb{R}$ and g is symmetric, the resulting Markov chain is never uniformly ergodic on $\mathbb{R}$. They also derived sufficient conditions for geometric ergodicity.

Tierney (1994) proposes an hybrid mix between the two previous algorithms by suggesting to use the transition density $g(y - a - b(x - a))$, i.e. to take

$$y_t = a + b(x^{(t)} - a) + \epsilon_t, \qquad \epsilon_t \sim g.$$

When $b < 0$, the random variables $x^{(t)}$ and $x^{(t+1)}$ may then be negatively correlated and this increases the mixing behavior of the chain (provided the value a is correctly chosen). Hastings (1970) considers an alternative to the uniform distribution on $[x^{(t)} - \delta, x^{(t)} + \delta]$ (see Example 1.2.1) by choosing the uniform distribution on $[-x^{(t)} - \delta, -x^{(t)} + \delta]$. The convergence of the empirical average (1.2) is then faster when $h(x) = x$, but the choice $a = 0$ is obviously central to this improvement. In general settings, b and a both need to be calibrated during preliminary runs of the algorithm.

A modification of the random walk Metropolis–Hastings algorithm has been proposed by Grenander and Miller (1994) and Phillips and Smith

(1996). It is based on a discretization of the *Langevin diffusion* associated with the stochastic differential equation

$$dX_t = dB_t + \frac{1}{2}\nabla \log f(X_t)dt, \tag{1.4}$$

where B_t is the standard *Brownian motion*. The derived Metropolis–Hastings algorithm is associated with the random walk like transition

$$x^{(t+1)} = x^{(t)} + \frac{\sigma^2}{2}\nabla \log f(x^{(t)}) + \sigma\varepsilon_t, \tag{1.5}$$

where $\varepsilon_t \sim \mathcal{N}_p(0, I_p)$ and σ^2 corresponds to the discretization step. While (1.5) may well be transient, a correction via the Metropolis–Hastings acceptance probability provides a theoretically sound algorithm which often enjoys better convergence[2] properties than the regular random walk Metropolis–Hastings algorithm. (See also Roberts and Rosenthal, 1997, for further results about the optimal choice of the scaling factor σ.)

Another direction has been opened by Green (1995) with his introduction of the *reversible jump Metropolis–Hastings algorithm*, which allows for *varying dimensions* of the parameter or, in other words, for setups where the dimension of the parameter space is a parameter itself. (Such settings are common in model choice for instance. See Richardson and Green, 1997, and Gruet, Philippe and Robert, 1998, for an illustration in the case of mixtures of distributions with an unknown number of parameters.) In such cases, the original algorithm [A_1] has to be adapted to deal with general state spaces and ratio of measures involving densities on spaces of different dimensions. Green (1995) proposes a generic approach, termed *reversible jump MCMC*, which uses a range of move types for changing dimensions, imposing that each corresponding transition kernel satisfies detailed balance. The move is accepted following a Metropolis–Hastings ratio where numerator and denominator are defined with respect to a common dominating measure, as detailed in Green (1995). The existence of such a measure is ensured if one has matched the degrees of freedom of joint variation of state space and proposal as dimension changes. In most cases, this amounts to defining a one-to-one correspondence between the largest dimension vector and the smallest dimension vector complemented by an appropriate vector of random variables.

Example 1.2.2 –**Step functions with random number of jumps**–
Similar to Green (1995), consider a distribution ℓ on $[0, 1]$ of the form

$$\ell(x) = \sum_{i=0}^{k}\omega_i \mathbb{I}_{[a_i, a_{i+1}]}(x),$$

---

[2] Note, however, that it requires an higher control on the stationary distribution f since the gradient $\nabla \log f$ must be available.

with

$$a_0 = 0, \quad a_{k+1} = 1 \quad \text{and} \quad \sum_{i=0}^{k} \omega_i(a_{i+1} - a_i) = 1 .$$

Assuming all the parameters are unknown (including k) and defining $p_i = \omega_i(a_{i+1} - a_i)$, consider the prior distribution

$$\pi(k, p^{(k)}, a^{(k)}) = \lambda^k e^{-\lambda} \frac{\Gamma((k+1)/2)p_0^{-1/2} \cdots p_k^{-1/2}}{\Gamma(1/2)^{k+1}} \mathbb{I}_{a_1 \leq \cdots \leq a_k},$$

which involves a Poisson distribution on k, $\mathcal{P}(\lambda)$, a uniform distribution on the ordered $\{a_1, \cdots, a_k\}$ and a Dirichlet distribution $D_{k+1}(1/2, \cdots, 1/2)$ on the weights p_i of the $\mathcal{U}_{[a_i, a_{i+1}]}$ distributions in ℓ.

Denoting the sample by $x = (x_1, \cdots, x_n)$, the posterior distribution on $k, \omega^{(k)}, a^{(k)}$ is

$$\pi(k, \omega^{(k)}, a^{(k)} | x) \quad \propto \quad \lambda^k \frac{\Gamma((k+1)/2)}{\Gamma(1/2)^{k+1}} \omega_0^{n_0 - 1/2} \cdots \omega_k^{n_k - 1/2}$$

$$\times \prod_{i=0}^{k} (a_{i+1} - a_i)^{-n_i} \mathbb{I}_{a_1 \leq \cdots \leq a_k},$$

where n_i is the number of observations between a_i and a_{i+1}.

Consider now that a model jump means a random birth or death of a step. When the current value of the dimension is $k + 1$, the *death move* is proposed with probability d_{k+1}:

1. Choose $1 \leq i \leq k+1$ at random with probability $1/(k+1)$.

2. Replace (ω_{i-1}, ω_i) with ω'_{i-1} such that

$$(a_{i+1} - a_{i-1})\omega'_i = (a_{i+1} - a_i)\omega_i + (a_i - a_{i-1})\omega_{i-1} ,$$

   leave other ω_j's unchanged, except for a decrease by 1 in the higher indices.

3. Delete a_i and decrease higher indices by 1 in the a_j's.

When the current value of the dimension is k, the *birth move* is proposed with probability b_k:

1. Generate $u_1 \sim \mathcal{U}_{[0,1]}$ and $u_2 \sim q(u_2)$.

2. Take $a_i^* = u_1$.

3. If $a_i \leq a_i^* \leq a_{i+1}$, propose new weights

$$\omega_{i-1}^* = \omega_i - \frac{u_2}{a_i^* - a_i}, \qquad \omega_i^* = \omega_i + \frac{u_2}{a_{i+1} - a_i^*}$$

   and replace ω_i with $(\omega_{i-1}^*, \omega_i^*)$.

In the above algorithm, the move is rejected if ω^*_{i-1} or ω^*_i in Step **3.** is negative.

Denoting by

$$\frac{\partial(\omega^{(k+1)}, a^{(k+1)})}{\partial(\omega^{(k)}, a^{(k)}, u_1, u_2)},$$

the matrix of the partial derivatives of the components of $(\omega^{(k+1)}, a^{(k+1)})$ with respect to the components of $(\omega^{(k)}, a^{(k)}, u_1, u_2)$, in the birth move, the Jacobian of the transform reduces to

$$\left| \frac{\partial(\omega^{(k+1)}, a^{(k+1)})}{\partial(\omega^{(k)}, a^{(k)}, u_1, u_2)} \right| = \frac{|a_{i+1} - a_i|}{|a^*_i - a_i||a_{i+1} - a^*_i|}$$

and the Metropolis–Hastings acceptance ratio for the birth move is

$$\rho_m = \min\left\{1, \frac{d_{k+1}}{(k+1)\,b_k\,q(u_2)} \frac{\pi(k+1, \omega^{(k+1)}, a^{(k+1)}|x)}{\pi(k, \omega^{(k)}, a^{(k)}|x)} \frac{|a_{i+1} - a_i|}{|a^*_i - a_i||a_{i+1} - a^*_i|}\right\}.$$

The Metropolis–Hastings acceptance ratio for the death move is obtained the opposite way. ‖

For an illustration in the setup of curve fitting, see Denison, Mallick and Smith (1998).

Except for the special case of the ARMS (which stands for *Accept-Reject Metropolis sampler*) algorithm where the instrumental density is automatically constructed from the objective distribution f (see Gilks, Best and Tan, 1993, for details), there are various degrees of arbitrariness involved in the choice of $q(\cdot|x)$. A comparison between different instrumental densities is quite delicate to undertake since it involves all possible uses of a sample from f. In particular, it is impossible to speak of an "optimal" choice of $q(\cdot|x)$.

At a humbler level, it is still possible to propose a calibration of the selected instrumental densities in terms of faster convergence to the stationary distribution. Consider then a parameterized family of conditional distributions $q(\cdot|x, \lambda)$. The choice of λ can be directed by a first calibration run of the Metropolis–Hastings algorithm, depending on the type of conditional distribution. If $q(\cdot|x, \lambda)$ is actually independent of x, λ should be chosen such that the acceptance probability in $[A_1]$ is as close to 1 as possible.[3]

Now, high acceptance probabilities are not particularly good for dependent algorithms, like random walk Metropolis–Hastings algorithms, because a high level of acceptance may indicate that the moves on the surface

---

[3] This obviously requires that the family of distributions $q(\cdot|\lambda)$ is close enough to the distribution of interest.

of f are not wide enough. In fact, if $x^{(t)}$ and y_t are close, in the sense that $f(x^{(t)})$ and $f(y_t)$ are similar, $[A_1]$ leads to accept y_t with probability

$$\min\left(\frac{f(y_t)}{f(x^{(t)})}, 1\right) \simeq 1 .$$

On the opposite, if the average acceptance rate is low, the successive values of $f(y_t)$ are often small compared with $f(x^{(t)})$, which means frequent excursions near the boundaries of the support of f. For such algorithms, an higher acceptance rate may thus mean a lower convergence speed, as the moves on the support of f are more limited. In the special case of multimodal densities, where modes are separated by almost zero probability zones, the negative consequences of limited moves on the support of f are quite obvious: although the acceptance rate is high for a distribution g with small variance, the probability to move from one mode to the other may be arbitrarily small.

As above, there is no general rule to decide whether an acceptance rate is too large or too low, but Gelman, Gilks and Roberts (1996) have developed a heuristic bound based on the example of the normal random walk and the approximation of $(x^{(t)})$ by a Langevin diffusion process. Their rule is to try to achieve an acceptance rate of about $1/4$ for high dimension models and about $1/2$ for models of dimension 1 or 2. Another version of this empirical rule is to choose *the scale factor of g equal to* $2.38/\sqrt{d}\ \Sigma$, where d is the model dimension and Σ the asymptotic variance of $x^{(t)}$ (which is usually an unknown quantity).

Example 1.2.3 Consider x_1, x_2, x_3 iid $\mathcal{C}(\theta, 1)$, a Cauchy distribution with location parameter θ and scale parameter 1, and $\pi(\theta) \propto 1$, the Lebesgue measure on $\mathbb{R}$. The posterior distribution on θ is

$$\pi(\theta|x_1, x_2, x_3) \propto [(1 + (\theta - x_1)^2)(1 + (\theta - x_2)^2)(1 + (\theta - x_3)^2)]^{-1}, \quad (1.6)$$

CAUCHY BENCHMARK

which is trimodal when x_1, x_2, x_3 are sufficiently different. (See for instance the case $x_1 = -8$, $x_2 = 8$, $x_3 = 17$). Since (1.6) has a large dispersion, we select a random walk Metropolis–Hastings algorithm based on a Cauchy distribution $\mathcal{C}(0, \sigma^2)$, where σ is to be calibrated.

Table 1.1 provides an estimation of the average acceptance probability ρ_σ, as well as the estimated variance of the estimators (1.2) of the expectations $\mathbb{E}^\pi[h_i(\theta)]$, for the functions[4]

$$h_1(\theta) = \theta, \quad h_2(\theta) = \left(\theta - \frac{17}{3}\right)^2 \quad \text{and} \quad h_3(\theta) = \mathbb{I}_{[4,8]}(\theta) ,$$

based on $20,000$ simulations. Note that there is not conclusive connection between the level of acceptance and the value of the corresponding variance,

---

[4]The value $17/3$ is chosen as the empirical average of the three observations.

but the various indicators are quite close and the asymptotic difference is presumably much smaller than the simulation variance.

TABLE 1.1. Performances of the algorithm $[A_1]$ associated with (1.6) for $x_1 = -8$, $x_1 = 8$ and $x_3 = 17$ and the random walk based on $\mathcal{C}(0, \sigma^2)$; ρ_σ is the approximated acceptance probability and the three last lines of the table provide the empirical variances for the evaluation of $\mathbb{E}^\pi[h_i(\theta)]$ (20,000 simulations).

σ	0.1	0.2	0.5	1.0	5.0	8.0	10.0	12.0
ρ_σ	0.991	0.985	0.969	0.951	0.893	0.890	0.891	0.895
h_1	41.41	44.24	44.63	43.76	42.59	42.12	42.92	42.94
h_2	0.035	0.038	0.035	0.036	0.036	0.036	0.035	0.036
h_3	0.230	0.228	0.227	0.226	0.228	0.229	0.228	0.230

An interesting feature of this experiment is that the probability of acceptance never goes below 0.88. Therefore, the goal of Gelman *et al.* (1996) cannot be achieved for this choice of instrumental distribution, for any value of σ.   ‖

1.3   The Gibbs sampler

A particular type of MCMC algorithm, called the Gibbs sampler (or Gibbs sampling), has been singled out for emerging independently in Geman and Geman (1984) (although it can also be traced back to Metropolis *et al.*, 1953) with a filiation closer to iterative algorithms like EM than to simulation (see, for instance, Tanner and Wong, 1987). Given f, assume there exists a density g such that f is the marginal density of g,

$$\int_Z g(x, z)\, dz = f(x),$$

and $p > 1$ such that the conditional densities of $g(y) = g(y_1, \cdots, y_p)$,

$$g_1(y_1|y_2, \cdots, y_p), g_2(y_2|y_1, y_3, \cdots, y_p), \cdots, g_p(y_p|y_1, \cdots, y_{p-1}),$$

can be simulated. The Gibbs sampler associated with this decomposition has the following transition from $y^{(t)}$ to $y^{(t+1)}$:

Simulate

1. $y_1^{(t+1)} \sim g_1(y_1|y_2^{(t)}, \cdots, y_p^{(t)})$;

2. $y_2^{(t+1)} \sim g_2(y_2|y_1^{(t+1)}, y_3^{(t)}, \cdots, y_p^{(t)})$, $[A_2]$

$\cdots$

p. $y_p^{(t+1)} \sim g_p(y_p|y_1^{(t+1)}, \cdots, y_{p-1}^{(t+1)})$.

Example 1.3.1 Consider the *Ising model*, where $(1 \leq i \leq I)$

$$f(s) \propto \exp\left\{-H\sum_i s_i - J\sum_{(i,j)\in\mathcal{N}} s_i s_j\right\}, \qquad s_i \in \{-1, 1\},$$

$\mathcal{N}$ being the *neighborhood relation* for the system (see Neal, 1993, for details on the Ising model). The full conditional distribution is then

$$\begin{aligned}
f(s_i|s_{j\neq i}) &= \frac{\exp\{-Hs_i - Js_i\sum_{j:(i,j)\in\mathcal{N}} s_j\}}{\exp\{-H - J\sum_{j:(i,j)\in\mathcal{N}} s_j\} + \exp\{H + J\sum_{j:(i,j)\in\mathcal{N}} s_j\}} \\
&= \frac{\exp\{-(H + \sum_{j:(i,j)\in\mathcal{N}} s_j)(s_i + 1)\}}{1 + \exp\{-2(H + \sum_{j:(i,j)\in\mathcal{N}} s_j)\}},
\end{aligned}$$

i.e. a logistic distribution on $(s_i + 1)/2$. The algorithm $[A_2]$ is thus easy to implement in this setup, the nodes i of the system being updated one by one. ‖

Some specific features of the Gibbs sampler are:

1. The acceptance rate of the Gibbs sampler is uniformly equal to 1. Every single simulated r.v. is thus accepted and the developments of the previous section on optimal acceptance rates are irrelevant.

2. The construction of the instrumental distribution requires precise information on f.

3. The Gibbs sampler is necessarily multidimensional, even though some components of the simulated vector may be artificial.

4. The Gibbs sampler does not work when the dimension of the vector space is variable.

While $f = g$ in Example 1.3.1, it is often necessary to complete x in $y = (x, z)$ and f in g to obtain manageable conditional g_i's, as in *data augmentation* setups (Tanner and Wong, 1987).

Example 1.3.2 Consider the distribution

$$f(\theta|\theta_0) \propto \frac{e^{-\theta^2/2}}{[1 + (\theta - \theta_0)^2]^\nu},$$

which corresponds to the posterior distribution of θ in either a Student or a normal setup (see Example 1.2.3); $f(\theta|\theta_0)$ can also be represented as the marginal distribution of

$$g(\theta, \eta) \propto e^{-\theta^2/2} e^{-[1+(\theta-\theta_0)^2]\eta/2} \eta^{\nu-1},$$

with conditional distributions

$$\eta \sim \mathcal{G}a\left(\nu, \frac{1 + (\theta - \theta_0)^2}{2}\right), \qquad \theta \sim \mathcal{N}\left(\frac{\theta_0\eta}{1+\eta}, \frac{1}{1+\eta}\right). \qquad ‖$$

The validity of the algorithm $[A_2]$ follows from its representation as the composition of p "elementary" Metropolis–Hastings algorithms, all with stationary distribution g and acceptance probability 1. In fact, the transition kernel for step i. is

$$q_i(y'|y) = \delta_{(y_1,\cdots,y_{i-1},y_{i+1},y_p)}(y_1',\cdots,y_{i-1}',y_{i+1}',y_p') \\ g_i(y_i'|y_1,\cdots,y_{i-1},y_{i+1},y_p) ,$$

where δ_y denotes the Dirac mass at y. The balance condition

$$\begin{aligned} g(y)q_i(y'|y) &= f(y)g_i(y_i'|y_1,\cdots,y_{i-1},y_{i+1},y_p) \\ &= g(y_1,\cdots,y_{i-1},y_i',y_{i+1},y_p)g_i(y_i|y_1,\cdots,y_{i-1},y_{i+1},y_p) \\ &= g(y')q_i(y|(y') \end{aligned}$$

then shows that g is the stationary distribution at each step. Moreover,

$$\begin{aligned} \frac{g(y')\,q_i(y|y')}{g(y)\,q_i(y'|y)} &= \frac{g(y')\,g_i(y_i|y_1,\cdots,y_{i-1},y_{i+1},y_p)}{g(y)\,g_i(y_i'|y_1,\cdots,y_{i-1},y_{i+1},y_p)} \\ &= \frac{g_i(y_i'|y_1,\cdots,y_{i-1},y_{i+1},y_p)\,g_i(y_i|y_1,\cdots,y_{i-1},y_{i+1},y_p)}{g_i(y_i|y_1,\cdots,y_{i-1},y_{i+1},y_p)\,g_i(y_i'|y_1,\cdots,y_{i-1},y_{i+1},y_p)} \\ &= 1. \end{aligned}$$

However, each of the p Metropolis–Hastings steps has very little to offer from a theoretical point of view, since it is not irreducible, being restricted to one component of y. That an irreducible chain can emerge from the composition is related with the *Hammersley–Clifford Theorem*, i.e. with the fact that a joint distribution can be reconstructed from its full conditionals. A sufficient condition, called *positivity condition*, is given in Besag (1974) (see also Hammersley, 1974): it imposes that the support of the joint distribution is the Cartesian product of the supports of the full conditionals. In this case (see Besag, 1974, 1994),

$$g(y_1,\cdots,y_p) \propto \prod_{j=1}^{p} \frac{g_{\ell_j}(y_{\ell_j}|y_{\ell_1},\cdots,y_{\ell_{j-1}},y_{\ell_{j+1}}',\cdots,y_{\ell_p}')}{g_{\ell_j}(y_{\ell_j}'|y_{\ell_1},\cdots,y_{\ell_{j-1}},y_{\ell_{j+1}}',\cdots,y_{\ell_p}')}$$

for every permutation ℓ on $\{1,2,\cdots,p\}$ and any fixed $y' \in \mathcal{Y}$. When the positivity condition holds, the algorithm $[A_2]$ produces an irreducible Markov chain. Besag (1994) and Hobert, Robert and Goutis (1996) propose some generalizations to the non-positive case, the former in terms of the Hammersley–Clifford Theorem and the later in terms of irreducibility of the associated Gibbs sampler.

As above, more advanced convergence properties like geometric or uniform geometric convergence cannot be guaranteed in general. For instance,

Tierney (1994) has established that a chain produced by $[A_2]$ is Harris recurrent when its transition kernel is absolutely continuous wrt the stationary measure.

Example 1.3.3 A model introduced by Gaver and O'Muircheartaigh (1987) is often used as a benchmark to compare different simulation strategies (see, e.g., Gelfand and Smith, 1990 or Tanner, 1996). It was originally proposed for the analysis of failures of nuclear pumps, as the dataset given in Table 1.2.

PUMP
BENCHMARK

TABLE 1.2. Number of failures and observation time for ten nuclear pumps (*Source :* Gaver and O'Muircheartaigh, 1987).

Pump	1	2	3	4	5	6	7	8	9	10
Failures	5	1	5	14	3	19	1	1	4	22
Time	94.32	15.72	62.88	125.76	5.24	31.44	1.05	1.05	2.10	10.48

The failures of the i-th pump are modeled according to a Poisson process with parameter λ_i $(1 \leq i \leq 10)$. For an observation time t_i, the number of failures p_i is thus distributed as a Poisson $\mathcal{P}(\lambda_i t_i)$ r.v. The corresponding prior distributions are chosen as

$$\lambda_i \overset{iid}{\sim} \mathcal{G}a(\alpha, \beta), \qquad \beta \sim \mathcal{G}a(\gamma, \delta) \qquad (1 \leq i \leq 10),$$

with $\alpha = 1.8$, $\gamma = 0.01$ and $\delta = 1$. The joint distribution is

$$
\begin{aligned}
&\pi(\lambda_1, \cdots, \lambda_{10}, \beta | t_1, \cdots, t_{10}, p_1, \cdots, p_{10}) \\
&\propto \prod_{i=1}^{10} \left\{ (\lambda_i t_i)^{p_i} \, e^{-\lambda_i t_i} \, \lambda_i^{\alpha-1} e^{-\beta \lambda_i} \right\} \beta^{10\alpha} \beta^{\gamma-1} e^{-\delta\beta} \\
&\propto \prod_{i=1}^{10} \left\{ \lambda_i^{p_i+\alpha-1} \, e^{-(t_i+\beta)\lambda_i} \right\} \beta^{10\alpha+\gamma-1} e^{-\delta\beta}
\end{aligned}
$$

and a possible decomposition of π in conditional distributions is

$$
\begin{aligned}
\lambda_i | \beta, t_i, p_i &\sim \mathcal{G}a(p_i + \alpha, t_i + \beta), \qquad (1 \leq i \leq 10) \\
\beta | \lambda_1, \cdots, \lambda_{10} &\sim \mathcal{G}a\left(\gamma + 10\alpha, \delta + \sum_{i=1}^{10} \lambda_i \right).
\end{aligned}
$$

The transition kernel on β associated with $[A_2]$ then satisfies

$$
K(\beta, \beta') = \int \frac{(\beta')^{\gamma+10\alpha-1}}{\Gamma(10\alpha+\gamma)} \left(\delta + \sum_{i=1}^{10} \lambda_i \right)^{\gamma+10\alpha} \exp\left\{ -\beta' \left(\delta + \sum_{i=1}^{10} \lambda_i \right) \right\}
$$

$$\times \prod_{i=1}^{10} \frac{(t_i + \beta)^{p_i + \alpha}}{\Gamma(p_i + \alpha)} \lambda_i^{p_i + \alpha - 1} \exp\{-(t_i + \beta)\lambda_i\} \, d\lambda_1 \ldots d\lambda_{10}$$

$$\geq \int \frac{\delta^{\gamma + 10\alpha}}{\Gamma(10\alpha + \gamma)} (\beta')^{\gamma + 10\alpha - 1} \exp\left\{-\beta'\left(\delta + \sum_{i=1}^{10} \lambda_i\right)\right\}$$

$$\times \prod_{i=1}^{10} \frac{(t_i + \beta)^{p_i + \alpha}}{\Gamma(p_i + \alpha)} \lambda_i^{p_i + \alpha - 1} \exp\{-(t_i + \beta)\lambda_i\} \, d\lambda_1 \ldots d\lambda_{10}$$

$$= \frac{\delta^{\gamma + 10\alpha}(\beta')^{\gamma + 10\alpha - 1}}{\Gamma(10\alpha + \gamma)} e^{-\delta\beta'} \prod_{i=1}^{10} \left(\frac{t_i + \beta}{t_i + \beta + \beta'}\right)^{p_i + \alpha}$$

$$\geq \frac{\delta^{\gamma + 10\alpha}(\beta')^{\gamma + 10\alpha - 1}}{\Gamma(10\alpha + \gamma)} e^{-\delta\beta'} \prod_{i=1}^{10} \left(\frac{t_i}{t_i + \beta'}\right)^{p_i + \alpha}$$

The kernel is therefore uniformly (in β) bounded from below, which implies that the corresponding chain $(\beta^{(t)})$ is uniformly ergodic. (It can also be shown that the dual chain of the $\lambda_i^{(t)}$'s is again uniformly ergodic.)   ‖

Example 1.3.4 (Example 1.3.2 cont.) The kernel on θ satisfies

$$K(\theta, \theta') = \int_0^\infty \sqrt{\frac{1 + \eta}{2\pi}} \exp\left\{-\left(\theta' - \frac{\theta_0 \eta}{1 + \eta}\right)^2 \frac{1 + \eta}{2}\right\} \left(\frac{1 + (\theta - \theta_0)^2}{2}\right)^\nu$$

$$\times \frac{\eta^{\nu - 1}}{\Gamma(\nu)} \exp\left\{\frac{-\eta}{2}(1 + (\theta - \theta_0)^2)\right\} d\eta,$$

$$\geq \int_0^\infty 2^{-\nu} \sqrt{\frac{1 + \eta}{2\pi}} \exp\left\{-\frac{1}{2}(\theta')^2 - \frac{1}{2}(\theta'^2 - 2\theta'\theta_0 + \theta_0^2)\eta\right\}$$

$$\times [1 + (\theta - \theta_0)^2]^\nu \frac{\eta^{\nu - 1}}{\Gamma(\nu)} \exp\left\{-\frac{1}{2}(1 + (\theta - \theta_0)^2)\eta\right\} d\eta$$

$$\geq \frac{e^{-(\theta')^2/2}}{\sqrt{2\pi}} \left[\frac{1 + (\theta - \theta_0)^2}{1 + (\theta - \theta_0)^2 + (\theta' - \theta_0)^2}\right]^\nu$$

$$\geq [1 + (\theta' - \theta_0)^2]^{-\nu} \frac{e^{-(\theta')^2/2}}{\sqrt{2\pi}}$$

and the chain $(\theta^{(t)})$ is thus uniformly ergodic.   ‖

The two examples above pertain to the special case of the Gibbs sampler where $p = 2$, which is called *data augmentation*, in analogy with the EM algorithm of Dempster, Laird and Rubin (1977) and stochastic restoration techniques (Celeux and Diebolt, 1985, Qian et Titterington, 1991, 1992), in *missing data setups*.

The duality between both components of the chain can be exploited to establish some properties of the joint chain or of one of the two subchains, as in Diebolt and Robert (1993, 1994). (See also Liu, Wong and Kong, 1994, 1995, for the notion of *interleaving chains*.) Section 1.5 and Chapter 3 point out the relevance of this feature in convergence control.

Example 1.3.5 Tanner and Wong (1987) consider the multinomial model

$$x \sim \mathcal{M}_5 \left(n; a_1\mu + b_1, a_2\mu + b_2, a_3\eta + b_3, a_4\eta + b_4, c(1 - \mu - \eta)\right),$$

with

$$0 \le a_1 + a_2 = a_3 + a_4 = 1 - \sum_{i=1}^{4} b_i = c \le 1,$$

where $a_i, b_i \ge 0$ are known. This model is derived from the completed model

$$y \sim \mathcal{M}_9 \left(n; a_1\mu, b_1, a_2\mu, b_2, a_3\eta, b_3, a_4\eta, b_4, c(1 - \mu - \eta)\right),$$

where some components have been aggregated:

$$x_1 = y_1 + y_2, \ x_2 = y_3 + y_4, \ x_3 = y_5 + y_6, \ x_4 = y_7 + y_8, \ x_5 = y_9.$$

A prior on (μ, η) is the Dirichlet distribution $\mathcal{D}(\alpha_1, \alpha_2, \alpha_3)$,

$$\pi(\mu, \eta) \propto \mu^{\alpha_1 - 1} \eta^{\alpha_2 - 1} (1 - \eta - \mu)^{\alpha_3 - 1}.$$

Set $z = (z_1, z_2, z_3, z_4) = (y_1, y_3, y_5, y_7)$. Then

$$
\begin{aligned}
\pi(\eta, \mu | x, z) &= \pi(\eta, \mu | y) \\
&\propto \mu^{\alpha_1 - 1} \eta^{\alpha_2 - 1} (1 - \eta - \mu)^{\alpha_3 - 1} \\
&\quad \times \mu^{z_1} \mu^{z_2} \eta^{z_3} \eta^{z_4} (1 - \eta - \mu)^{x_5} \\
&= \mu^{z_1 + z_2 + \alpha_1 - 1} \eta^{z_3 + z_4 + \alpha_2 - 1} (1 - \eta - \mu)^{x_5 + \alpha_3 - 1}.
\end{aligned}
$$

Therefore,

$$(\mu, \eta, 1 - \mu - \eta) | x, z \sim \mathcal{D}(z_1 + z_2 + \alpha_1, z_3 + z_4 + \alpha_2, x_5 + \alpha_3).$$

Moreover,

$$
\begin{aligned}
z_i | x, \mu, \eta &\sim \mathcal{B}\left(x_i, \frac{a_i \mu}{a_i \mu + b_i}\right), &(i = 1, 2) \\
z_i | x, \mu, \eta &\sim \mathcal{B}\left(x_i, \frac{a_i \eta}{a_i \eta + b_i}\right), &(i = 3, 4).
\end{aligned}
$$

Both conditional distributions $g_1(\theta | y)$ and $g_2(z | x, \theta)$ are thus available and the distribution on the z_i's has furthermore a finite support.   ‖

The special features of the Gibbs sampler sometimes allow for alternatives to the empirical averages (1.2) which exploit the conditional nature of the simulations. Gelfand and Smith (1990) have thus introduced a conditioning method called *Rao-Blackwellization* by reference to the Rao-Blackwell Theorem (see Lehmann and Casella, 1998). *Rao-Blackwellization* replaces

$$\delta_0 = \frac{1}{T} \sum_{t=1}^{T} h\left(y_1^{(t)}\right)$$

with

$$\delta_{rb} = \frac{1}{T} \sum_{t=1}^{T} \mathbb{E} \left[h(y_1) | y_2^{(t)}, \cdots, y_p^{(t)} \right] ,$$

when the conditional expectation can be easily computed. Both estimators are unbiased and converging to $\mathbb{E}_g[h(y_1)]$. Domination of δ_0 by δ_{rb} does not follow from the usual Rao-Blackwell Theorem though, because the $y_i^{(t)}$ are dependent. Liu, Wong and Kong (1994) show that it holds for all interleaving chains and, in particular, for Data Augmentation (see also Geyer, 1995, for necessary conditions). Chapter 3 will come back to the Rao-Blackwellization as a convergence control device.

Compared with a general Metropolis–Hastings algorithm such as independent or random walk Metropolis–Hastings schemes, the Gibbs sampler has both advantages and drawbacks. On the one hand, the Gibbs sampler is directly based on the true density f while an Metropolis–Hastings algorithm is at best founded on an approximation of this distribution and thus offers more chances of misspecification. On the other hand, the Gibbs sampler is a composition of very poor Metropolis–Hastings algorithms which only update one component at a time and are thus not irreducible. Moreover, the decomposition of y in components $y_1, \ldots, y_p$ may be unrelated to the directions of importance for the density f and may thus induce a very slow exploration of the support of f. Imagine for instance a density in $\mathbb{R}^2$ whose support is mostly concentrated on both diagonals of $\mathbb{R}^2$. Excursions from one branch of f to another are likely to be very rare if the current system of coordinates is selected. (Hills and Smith, 1992, 1993, propose examples where an ill-chosen parameterization dramatically increases convergence times.)

The defects of Metropolis–Hastings algorithms usually stem from a lack of information on f and they can sometimes be recovered by calibrating parameters of the instrumental distributions. They are in any case quite useful to explore the rough features of f, the Gibbs sampler being by contrast more adequate to provide the finer details. To take advantage of both perspectives, *hybrid* strategies which involve both Gibbs and Metropolis–Hastings steps have been proposed by Tierney (1994) and others. Given kernels $K_1, \cdots, K_n$ and a probability distribution $(\alpha_1, \cdots, \alpha_n)$, the *mixture* of $K_1, K_2 \ldots$ and K_n is the kernel

$$\tilde{K} = \alpha_1 K_1 + \cdots + \alpha_n K_n .$$

Similarly, the *cycle* of $K_1, K_2, \ldots$ and K_n is the kernel

$$K^* = K_1 \circ \cdots \circ K_n.$$

This definition includes cases where a Metropolis–Hastings step is chosen every m Gibbs steps, a strategy which is recommended in the case of multimodal densities. It also includes random ordering of the p steps of $[A_2]$ where the conditional distributions are simulated in a random sequence. From a theoretical point of view, an interesting feature of these hybrid algorithms is that good convergence properties of one of the kernels K_i usually transfer to the resulting kernel for mixtures and often for cycles (see Tierney, 1994).

There is also a use for hybrid MCMC algorithms at a more elementary level, namely when some components of $[A_2]$ cannot be easily simulated. As suggested by Müller (1991, 1993), each step i. where the direct simulation from $g_i(y_i|y_j, j \neq i)$ is impossible can be replaced by a simulation from an instrumental distribution q_i :

i.1. Simulate $\tilde{y}_i \sim q_i(y_i|y_1^{(t+1)}, \cdots, y_i^{(t)}, y_{i+1}^{(t)}, \cdots, y_p^{(t)})$

i.2. Take

$$y_1^{(t+1)} = \begin{cases} y_i^{(t)} & \text{with probability } 1 - \rho, \\ \tilde{y}_i & \text{with probability } \rho, \end{cases}$$

$$[A_3]$$

where

$$\rho = 1 \wedge \left\{ \frac{g_i(\tilde{y}_i|y_1^{(t+1)}, \cdots, y_p^{(t)})}{q_i(\tilde{y}_i|y_1^{(t+1)}, \cdots, y_i^{(t)}, \cdots, y_p^{(t)})} \middle/ \frac{g_i(y_i^{(t)}|y_1^{(t+1)}, \cdots, y_p^{(t)})}{q_i(y_i^{(t)}|y_1^{(t+1)}, \cdots, \tilde{y}_i, \cdots, y_p^{(t)})} \right\} .$$

Note that this Metropolis–Hastings step is only used *once* in $[A_2]$. Therefore, there is no convergence issue at this level, in the sense that the algorithm does not aim at approximating $g_i(y_i|y_j, j \neq i)$, but replaces this instrumental distribution, associated with an acceptance probability of 1, with another instrumental distribution. Indeed, the hybrid step is quite valid from an MCMC point of view since g is still the stationary distribution of the chain. Besides, the conditional distribution $g_i(y_i|y_j, j \neq i)$ is *not* the distribution of interest and does not need to be well approximated.

Müller (1993) notes that the acceptance step can be transferred at the end of the p conditional simulations, instead of being reproduced at each hybrid replacement. The appeal of this modification is to express the algorithm as a genuine Metropolis–Hastings algorithm.

1.4 Perfect sampling

In some settings, it is quite important to know when stationarity is "attained" or, more exactly, to start the Markov chain in its stationary regime, $x^{(0)} \sim f(x)$, in order to avoid the bias caused by the initial value/distribution and to concentrate on either the determination of an acceptable batch size k, so that $x^{(0)}, x^{(k)}, x^{(2k)}, \ldots$ are nearly independent, or the accuracy of an ergodic average (1.2). In other cases, one needs to know, as put by Fill (1998a), *"how long is long enough"*, in order to evaluate the necessary computing time or the mixing properties of the chain.

Following Propp and Wilson (1996), several authors have proposed devices to sample directly from the stationary distribution f at varying computational costs and for specific distributions and transitions. The denomination of *perfect sampling* for such techniques was coined by Kendall (1996), replacing the *exact sampling* terminology of Propp and Wilson (1996). The main bulk of the work, so far, deals with finite state spaces; this is due, for one thing, to the higher simplicity of these spaces and, for another, to statistical physics motivations related to the Ising model (see Example 1.3.1). The appeal of these methods for mainstream statistical problems is yet unclear, but Murdoch and Green (1998) have shown that some standard examples like the pump benchmark (Example 1.3.3) do allow for *perfect sampling* in continuous settings. Moreover, when the Duality Principle of §1.5 applies, the stationarity of the finite chain obviously transfers to the dual chain, even if the latter is continuous. (We will see in Chapter 6 the case of a hidden Markov model where this setting occurs.)

The method proposed by Propp and Wilson (1996) is called *coupling from the past* (CFTP). It applies to chains on a finite state space $\mathcal{X}$, represented by $\{1, \ldots, p\}$, and it couples p chains initialized at the p possible states in $\mathcal{X}$ farther and farther back in the past till all chains take the same value (or *coalesce*) at time 0. Various levels of *coupling* of the chains (see §2.4) can be applied, in order to increase the convergence speed, ensuring for instance that the chains are indistinguishable from the time they have met.

Given a transition matrix $\mathbb{P}$ with probabilities p_{uv}, $u, v \in \mathcal{X}$, define the associated cumulative probabilities q_{uv}, $u, v \in \mathcal{X}$, as

$$q_{uv} = \sum_{w=1}^{v} p_{uw}.$$

The CFTP technique goes back in time by constructing functions (or *maps*) $f_{-1}, f_{-2}, \cdots, f_{-N}$ from $\mathcal{X}$ to $\mathcal{X}$ which represent the transitions at times $-1, -2, \ldots, -N$. They depend on uniform rv's $\omega_1, \omega_2, \ldots, \omega_N$ in the following way $(u \in \mathcal{X})$:

$$f_{-t}(u) = \min\{v; \omega_t \leq q_{uv}\}. \tag{1.7}$$

Therefore, $P(f_{-t}(u) = v) = p_{uv}$, $u, v \in \mathcal{X}$, and the construction respects (marginally) the transition. The functions F_{-t} are then deduced by $(u \in \mathcal{X})$

$$F_{-t}(u) = F_{-(t-1)}(f_{-t}(u)),$$

with $F_0 = id_{\mathcal{X}}$. They represent realizations of the N-fold chain, i.e. $F_{-N}(u) = x^{(0)}$ when $x^{(-N)} = u$. The (past) horizon N is then increased until F_{-N} is constant, that is when $x^{(0)}$ is the same for every starting value $x^{(-N)}$. Note that N is thus a (backward) stopping time which depends on the sequence of uniforms $\omega_1, \ldots$

Lemma 1.4.1 *When the chain is irreducible and aperiodic, coalescence takes place with probability one and the random variable $x^{(0)}$ produced by the CFTP method is distributed from the stationary distribution f.*

Proof. For simplicity's sake, we assume that the chain is strongly aperiodic, namely that $p_{uv} > 0$ for every u, v. Therefore, the probability that f_{-t} is constant satisfies

$$P(f_{-t} = f_{-t}(1)) = \sum_v \prod_u p_{uv} = \alpha > 0 .$$

Besides, the f_{-t}'s are independent, being functions of the ω_{-t}'s, and the Borel–Cantelli lemma implies that f_{-t} is constant infinitely often. Therefore, there exists $\hat{N}$ such that $F_{-\hat{N}}$ is constant. Since $F_{-\hat{N}-1} = x^{(0)}$ satisfies

$$
\begin{aligned}
P(x^{(1)} = x) &= \sum_y P(x^{(1)} = x | x^{(0)} = y) P(x^{(0)} = y) \\
&= \sum_y p_{yx} P(x^{(0)} = y),
\end{aligned}
$$

and $x^{(1)}$ and $x^{(0)}$ have then the same distribution, it is necessarily the stationary distribution of interest.   □□

Propp and Wilson (1996) suggest to update the maps at dates -1, -2, -4, ..., till coalescence as this updating scheme has nearly optimal properties. An important remark of these authors (which has been emphasized in many of the following papers) is that things are sped up when a *"monotonicity"* structure is available, namely when there exist a (stochastically) larger state x_1 and a (stochastically) smaller state x_0. In this case, it is sufficient to consider chains starting from the two states x_0 and x_1 till they coincide at time 0 since $F_{-N}(x_0) = F_{-N}(x_1)$ implies coalescence, given that all the intermediary paths are located between the two extreme cases. Although this appears as a strong requirement on the model structure, we will see in Chapter 6 a statistical setup where monotonicity occurs. Note also that the validity of the method extends to data augmentation settings (see §1.3), even when the second chain in the structure has a continuous support. The construction of the map is then more complex than in (1.7), since it implies generating the second chain ($\theta^{(-t)}$), but, once this generation is operated, the transition from x to y at time $-t$ is also determined and the generation works as a black box.

Despite the general convergence result of Lemma 1.4.1 that coalescence requires a finite number of backward iterations, there are still problems with the CFTP algorithm, in that the running time is nonetheless unbounded and that aborting long runs does create bias. Fill (1998a,b) proposes an alternative algorithm which can be interrupted while preserving the central feature of CFTP. It is however restricted to monotone and discrete settings.

Murdoch and Green's (1998) extension to the continuous case is based on renewal theory (see §4.2 for a short introduction to renewal theory), either for uniformly ergodic chains or for Metropolis-Hastings algorithms with atoms, including, paradoxically, the independent case (see §4.2.2). It is also a CFTP technique, formally using a continuum of chains starting from all possible values in $\mathcal{X}$ at time $-1, -2, -4, \ldots$ till they all coalesce into a single chain at time[5] 0. The authors propose several techniques to replace the continuum of chains by a finite number of (coupled) chains (see also Green and Murdoch, 1998). For instance, in the special case the transition kernel satisfies *Doeblin's condition*,

$$K(x, y) \geq r(x) , \quad x, y \in \mathcal{X} , \tag{1.8}$$

which is equivalent to uniform ergodicity (see, e.g., Example 1.3.3), the continuum of Markov chains coalesces into a single chain at each time with probability

$$\rho = \int r(x) dx .$$

That the resulting random variable at time 0 is distributed from the stationary distribution f follows from a coupling argument: at each time, a generation from the bounding probability $\rho^{-1} r(x)$ occurs with probability ρ and the *same uniform variable* is used to decide whether this generation occurs for all chains. The validity result of Lemma 1.4.1 thus extends to this setting. In practice, this means that the single chain is started at a random moment $-N$ in the past from the bounding probability $\rho^{-1} r(x)$ with $N \sim \mathcal{G}eo(\rho)$.

Example 1.4.1 As shown in Example 1.3.3, the chain $(\beta^{(t)})$ produced by the Gibbs sampler is uniformly ergodic, with bounding measure

PUMP
BENCHMARK

$$r(\beta') = \frac{\delta^{\gamma + 10\alpha} (\beta')^{\gamma + 10\alpha - 1}}{\Gamma(10\alpha + \gamma)} e^{-\delta\beta'} \prod_{i=1}^{10} \left(\frac{t_i}{t_i + \beta'} \right)^{p_i + \alpha}$$

Since the normalized version $r(\beta)/\varrho$ is not a standard distribution, the implementation of the CFTP algorithm requires to determine the value of ρ and to simulate from this distribution.[6] The first aspect can be solved

---

[5] The fact that the chain is always considered at time 0 and not at the first time all chains coalesce is important, because it eliminates the bias created by the stopping (meeting) time.

[6] Note that this is not the solution advocated in Murdoch and Green, 1998.

by using the Riemann integration techniques of Philippe (1997a) (see §3.3) or by one of the normalizing constant estimation technique given in Chen and Shao (1997). The simulation issue can be addressed through an accept-reject algorithm based on the gamma $\mathcal{G}a(\gamma + 10\alpha, \delta)$.   ‖

When (1.8) does not hold, $\mathcal{X}$ can be replaced (in theory, see Meyn and Tweedie, 1993) by a partition of m sets $\mathcal{A}_i$ such that

$$K(y|x) \geq r_i(x), \quad y \in \mathcal{X}, \quad x \in \mathcal{A}_i,$$

as we will see in §4.2. Murdoch and Green (1998) then propose to start m chains started the bounding distributions $\rho_i^{-1} r_i$ at a random time in the past and to couple them till they coalesce, using the renewal decomposition

$$K(y|x) = \rho_i[\rho_i^{-1} r_i(x)] + (1 - \rho_i)\left[\frac{K(y|x) - r_i(x)}{1 - \rho_i}\right],$$

i.e. simulating with probability ρ_i from $\rho_i^{-1} r_i(x)$ at each time. As in the original CFTP algorithm, the sequence of uniform variables used in the selection (between $\rho_i^{-1} r_i(x)$ and the complement) and the generation from the selected component is fixed and common to all chains.

1.5   Convergence results from a Duality Principle

We now focus on a specific property, called the *Duality Principle*, which relates different components of an (MCMC) Markov chain in terms of the-oretical convergence properties, as it is strongly connected with control purposes, since this principle shows that convergence performances can be deduced from a subchain of the global Markov chain, even when this sub-chain has a finite support.

Diebolt and Robert (1993, 1994) introduced the *Duality Principle* for the estimation of mixtures, which is presented in §4.4, but let us start with a simple example.

Example 1.5.1 The Gibbs algorithm is obtained by completion of the
MULTINOMIAL data from x to $y = (x, z)$ as
BENCHMARK

1. Simulate $z = (z_1, z_2, z_3, z_4) = (y_1, y_3, y_5, y_7)$ by

$$z_i \sim \mathcal{B}\left(x_i, \frac{a_i\mu}{a_i\mu + b_i}\right) \ (i = 1, 2), \qquad z_i \sim \mathcal{B}\left(x_i, \frac{a_i\eta}{a_i\eta + b_i}\right) \ (i = 3, 4).$$

2. Simulate $[A_4]$

$$(\mu, \eta) \sim \mathcal{D}(1/2 + z_1 + z_2, 1/2 + z_3 + z_4, 1/2 + x_5).$$

The chain $(z^{(t)})$ is generated on a finite state space of cardinal $(x_1 + 1) \times (x_2 + 1) \times (x_3 + 1) \times (x_4 + 1)$, while $(\mu^{(t)}, \eta^{(t)})$ can be seen as the *dual chain* of $(z^{(t)})$, in the sense that $z^{(t)}$ is generated conditionally on $(\mu^{(t-1)}, \eta^{(t-1)})$, while $(\mu^{(t)}, \eta^{(t)})$ is generated conditionally on $z^{(t)}$. The continuous chain $(\mu^{(t)}, \eta^{(t)})$ is therefore a (random) transform of the finite state space chain $(z^{(t)})$. ‖

Example 1.5.2 Consider a *capture-recapture model* where the size N of an overall population is unknown and where each individual in this population has a probability p to be captured for each capture experiment, whatever his past history and the history of the other individuals in the population. (Individuals captured in a given capture experiment are released for the next experiment.) For two successive experiments, a sufficient statistic is (n_{11}, n_{12}, n_{21}) where n_{11} is the number of individuals captured twice, n_{12} the number of individuals captured only the first time and n_{21} the number of individuals captured only the second time. The likelihood of this so-called *uniform* model (Castledine, 1981; Wolter, 1986; George and Robert, 1992) is

$$\ell(N, p | n_{11}, n_{12}, n_{21}) \propto \begin{pmatrix} N \\ n_{11} \quad n_{12} \quad n_{21} \end{pmatrix} p^{n_0}(1-p)^{N-n_0},$$

with $n_0 = n_{12} + n_{21} + 2n_{11}$, the total number of captures. This likelihood is factorizing through n_0 and $n_t = n_{12} + n_{11} + n_{21}$, total number of different captured individuals. If $\pi(p, N)$ corresponds to a Poisson $\mathcal{P}(\lambda)$ distribution on N and to a uniform $\mathcal{U}_{[0,1]}$ distribution on p, the posterior distribution of (p, N) is

$$\pi(p, N | n_0, n_t) \propto \frac{N!}{(N-n_t)!} \, p^{n_0}(1-p)^{N-n_0} \, \frac{e^{-\lambda}\lambda^N}{N!}$$

which implies that

$$\begin{aligned} (N - n_t)|p, n_0, n_t &\sim \mathcal{P}(\lambda) \,, \\ p|N, n_0, n_t &\sim \mathcal{B}e(n_0 + 1, N - n_0 + 1). \end{aligned}$$

Therefore, Data Augmentation applies in this setup, leading to a chain $(p^{(t)})$ generated conditionally on a Markov chain $(N^{(t)})$ with discrete support. (Note that direct computations are available for the uniform model, see George and Robert, 1992.) ‖

In many MCMC setups similar to Examples 1.5.1 and 1.5.2, the algorithm produces several chains in parallel. This is particularly true of *Data Augmentation* (Tanner and Wong (1987), as in the two previous examples, and of general Gibbs sampling (Gelfand and Smith, 1990). In some cases, as in the extension of Example 1.5.2 to temporal capture-recapture models where the probability of capture changes with time (see Dupuis, 1995, or Robert, 1996c) or more generally in hidden mixture setups (see §3.4) or even more globally in Gibbs sampling algorithms with more than two conditional distributions, the chains of interest $(\theta^{(t)})$ are not necessarily Markov chains but the Duality Principle we now introduce shows this is not really a concern. In a somehow paradoxical way, this principle asserts that it is not always appropriate to study directly the chain of interest and that it may be more interesting to borrow strength from the simpler of the chains under study.

More precisely, the *Duality Principle* leading to Theorems 1.5.1 and 1.5.2 states that in cases where the chain $(\theta^{(t)})$ is derived from a second chain $(z^{(t)})$ by simulation from $\pi(\theta|z)$, the properties of the chain $(\theta^{(t)})$, *whether it is a Markov chain or not*, can be gathered from those of the chain $(z^{(t)})$. In this setup, $z^{(t)}$ is simulated according to the conditional distribution $f(z|\theta^{(t-1)}, z^{(t)})$. In the particular case $z^{(t)}$ has a finite support, one can fully appreciate the impact of this principle, even though some finite setups such as the Ising model (see Neal, 1993), or even the mixture example of §3.4, may lead to difficulties.

Theorem 1.5.1 *If the chain $(z^{(t)})$ is ergodic with stationary distribution $\tilde{f}$ (respectively geometrically ergodic with rate ϱ), the chain $(\theta^{(t)})$ derived by $\theta^{(t)} \sim \pi(\theta|z^{(t)})$ is ergodic (geometrically ergodic) for every conditional distribution $\pi(\cdot|z)$ and its stationary distribution is*

$$\tilde{\pi}(\theta) = \int \pi(\theta|z)\tilde{f}(z)dz.$$

Proof. The transition kernel associated to the chain $(z^{(t)})$ is

$$K(z, z') = \int \pi(\theta|z)f(z'|\theta, z)d\theta.$$

If f^t is the marginal density of $z^{(t)}$, $\pi^t(\theta) = \int \pi(\theta|z)f^t(z)dz$ is the marginal density of $\theta^{(t)}$ and

$$
\begin{aligned}
||\pi^t - \tilde{\pi}||_{TV} &= 1/2 \left| \int_{Z \times \Theta} \pi(\theta|z)(f^t(z) - \tilde{f}(z))dz d\theta \right| \\
&\leq ||f^t - \tilde{f}||_{TV},
\end{aligned}
$$

where $|| \cdot ||$ denotes the *total variation norm*, which is also half the L_1 distance. Therefore, $(\theta^{(t)})$ converges to $\tilde{\pi}$ for every possible starting point and the chain is ergodic whenever $(z^{(t)})$ is ergodic. The same transfer applies for geometric ergodicity. Note that the inequalities

$$||f^{t+1} - \tilde{f}||_{TV} \leq ||\pi^t - \tilde{\pi}||_{TV} \leq ||f^t - \tilde{f}||_{TV}$$

imply that the same geometric rate ϱ applies to both chains.   □□

The *Duality Principle* corresponding to this result then states that general convergence properties of the chain $(z^{(t)})$ can be extended not only to deterministic transforms $(h(z^{(t)}))$, but also to *random transforms*. For instance, once $(z^{(t)})$ has converged to its stationary distribution, the transform of $(z^{(t)})$ by the transition $\pi(\cdot|z)$ is then obviously stationary. Also, the fact that the same geometric rate applies to both chains can be used for convergence control since ρ is usually easier to estimate for finite state space Markov chains (see, e.g., Saloff-Coste and Diaconis, 1993, and Roberts and Tweedie, 1996). In an informal way, this principle recalls the mixture technique used in simulation, where a r.v. θ is generated as the component of a vector (θ, z) which is easier to simulate.

In the special case when $(\theta^{(t)})$ is a Markov chain, e.g. when $z^{(t)} \sim f(z^{(t)}|\theta^{(t-1)})$, which corresponds to Data Augmentation, α-mixing and β-mixing properties (see Bradley, 1986) also transfer from $(z^{(t)})$ to $(\theta^{(t)})$. This may have bearings on the Central Limit Theorem for transforms of $(\theta^{(t)})$ (see Robert, 1995).

Theorem 1.5.2 *If the chain $(z^{(t)})$ is α-mixing (respectively β-mixing), the chain $(\theta^{(t)})$ is also α-mixing (β-mixing).*

Proof. Consider the following representation of the α-mixing coefficients

$$\alpha_\theta(t) = \sup_{||h||_\infty < 1} \int_\Theta \left| \int_\Theta h(\theta)(\pi^t(\theta|\theta_0) - \tilde\pi(\theta))d\theta \right| \tilde\pi(\theta_0)d\theta_0.$$

Then

$$
\begin{aligned}
\alpha_\theta(t) &\leq \sup_{||h||_\infty < 1} \int_\Theta \left| \int_Z \int_\Theta h(\theta)\pi(\theta|z)d\theta (f^t(z|\theta_0) - \tilde f(z))dz \right| \tilde\pi(\theta_0)d\theta_0 \\
&\leq \sup_{||g||_\infty < 1} \int_\Theta \int_Z \left| \int_Z g(z)(f^{t-1}(z|z_1) - \tilde f(z))dz \right| f(z_1|\theta_0)dz_1 \tilde\pi(\theta_0)d\theta_0 \\
&= \sup_{||g||_\infty < 1} \int_Z \left| \int_Z g(z)(f^{t-1}(z|z_1) - \tilde f(z))dz \right| \tilde f(z_1)dz_1 = \alpha_z(t-1).
\end{aligned}
$$

Similarly, since (Davydov, 1973)

$$\beta_\theta(t) = \int_\Theta \int_\Theta |\pi^t(\theta|\theta_0) - \tilde\pi(\theta)|d\theta \tilde\pi(\theta_0)d\theta_0,$$

we get

$$
\begin{aligned}
\beta_\theta(t) &\leq \int_\Theta \int_Z |f^t(z|\theta_0) - \tilde f(z)|dz \tilde\pi(\theta_0)d\theta_0 \\
&\leq \int_\Theta \int_Z \int_Z |f^t(z|z_1) - \tilde f(z)|dz f(z_1|\theta_0)\tilde\pi(\theta_0)d\theta_0 \\
&= \int_Z \int_Z |f^{t-1}(z|z_0) - \tilde f(z)|dz \tilde f(z_0)dz_0 = \beta_z(t-1) ,
\end{aligned}
$$

and these inequalities complete the proof of Theorem 1.5.2.   □□

This correspondence between the probabilistic properties of both chains is directly of interest, since it allows for an assessment of convergence on the simpler (finite state space) chain using standard tools of finite Markov chain theory and for a free derivation of convergence assessments on the general chain $(\theta^{(t)})$. This derivation sounds slightly paradoxical since $(z^{(t)})$ is finite and $(\theta^{(t)})$ is continuous while they converge exactly at the same speed but one can see the limitations of this kind of reasoning, since it leads to true paradoxes related to the uncountable nature of continuous distributions, like a "requirement" to visit each point of the continuous state space!

The Duality Principle may then help in the assessment that the Central Limit Theorem holds. In fact, the sufficient conditions described in Meyn and Tweedie (1993) (see also Tierney, 1994) have only to be checked for the chain $(z^{(t)})$ for the Central Limit Theorem to apply to the chain $(\theta^{(t)})$. When $(z^{(t)})$ is a finite state space Markov chain, such verification is often straightforward. For instance, it follows from Billingsley (1968) that $(z^{(t)})$ is geometrically ergodic and even φ-mixing, under irreducibility and aperiodicity of the kernel. At this stage, the Central Limit Theorem does not appear much as a convergence control device, given its formal nature, but Chapter 5 will hopefully demonstrates that there are some genuine consequences of the Central Limit Theorem in convergence control.

The mixture estimation example described in §3.4, which was instrumental in the derivation of the Duality Principle by Diebolt and Robert (1993), is another example where a finite state space Markov chain dictates the type of convergence of a whole continuous chain, but it is far from being the only setup where the Duality Principle applies with practical consequences. Tanner (1996) provides on the contrary many examples in this spirit (see also Robert, 1996c, Chapter 7, for missing data illustrations). For instance, *grouping* as in Heitjan and Rubin (1991) is another type of missing data structure where the Duality Principle can strengthen the convergence study, although the following example does not relate to finite state space Markov chains.

Example 1.5.3 Consider p random variables $y_1, \ldots, y_p \sim \mathcal{E}xp(\theta)$ which are grouped into classes according to Bernoulli random variables $g_i \sim \mathcal{B}(\Phi(\gamma_1 - \gamma_2 y_i))$ as follows

$$x_i = \begin{cases} [y_i/a] & \text{if } g_i = 0, \\ [y_i/b] & \text{if } g_i = 1, \end{cases} \quad (1 \leq i \leq p)$$

where a, b, γ_1 and γ_2 are known, and Φ is the normal cdf. Heitjan and Rubin (1991) provide a justification for this model through round-up errors in surveys. The observations x_i can be completed by the missing data (y_i, g_i)

and

$$f(y_i, g_i | x_i, \theta) \quad \propto \quad \theta e^{-\theta y_i} \big\{ \mathbb{I}_{[ax_i, a(x_i+1)]}(y_i) \mathbb{I}_{g_i=0} [1 - \Phi(\gamma_1 - \gamma_2 y_i)] + \\ \mathbb{I}_{[bx_i, b(x_i+1)]}(y_i) \mathbb{I}_{g_i=1} \Phi(\gamma_1 - \gamma_2 y_i) \big\}.$$

If the prior distribution on θ is a $\mathcal{G}a(\alpha, \beta)$ distribution, a Gibbs algorithm for the simulation of the posterior distribution of θ is to consider the Markov chain $(z^{(t)}, \theta^{(t)})$ with transition step

1. Simulate $z^{(t)} = (y_1^{(t)}, g_1^{(t)}, \ldots, y_p^{(t)}, g_p^{(t)})$ **by** $(1 \leq i \leq p)$

$$g_i^{(t)} \quad \sim \quad \mathcal{B}(1, \Phi(\gamma_1 - \gamma_2 y_i^{(t-1)})),$$
$$y_i^{(t)} | g_i^{(t)} \quad \sim \quad \theta^{(t-1)} e^{-\theta^{(t-1)} y} \big\{ \mathbb{I}_{[ax_i, a(x_i+1)]}(y_i) \mathbb{I}_{g_i^{(t)}=0}$$
$$+ \mathbb{I}_{[bx_i, b(x_i+1)]}(y_i) \mathbb{I}_{g_i^{(t)}=1} \big\}.$$

2. Simulate $\hspace{10cm}$ $[A_5]$

$$\theta^{(t)} \sim \mathcal{G}a\left(\alpha + p, \beta + \sum_{i=1}^{p} y_i^{(t)}\right).$$

In this case, the missing data (y_i, g_i) $(1 \leq i \leq p)$ has a compact support and the chain $(z^{(t)})$ is φ-mixing. In particular, the Central Limit Theorem applies to both $(z^{(t)})$ and $(\theta^{(t)})$. $\hspace{4cm}$ ‖

2
Convergence Control of MCMC Algorithms

Christian P. Robert
Dominique Cellier

2.1 Introduction

There is an obvious difference between the theoretical guarantee that f is the stationary distribution of a Markov chain $(x^{(t)})$ and the practical requirement that (1.2) is close enough to (1.1). It is thus necessary to develop diagnostic tools towards the latter goal, namely convergence control.[1] While control is the topic of this book, we first present in this chapter some of the usual methods, before embarking upon the description of new control methods. The reader is referred to the survey papers of Brooks (1998), Brooks and Roberts (1998) and Cowles and Carlin (1996), as well as to Robert and Casella (1998) and Gelfand and Smith (1998) for details.

We first distinguish between *single chain* (§2.2) and *parallel chain* (§2.3) control methods because both their motivations and purposes differ. The former are indeed usually oriented towards a control of the convergence of (1.2) to (1.1) for arbitrary functions h or, in other words, towards an assessment of the mixing properties of the chain $(x^{(t)})$, that is of the speed of exploration of the support of f. Besides, they usually allow for an "on-line" processing of the simulation output. The later methods require on the contrary a tailored implementation, with several runs of the algorithm and a preliminary selection of the initial distribution. Moreover, they are closer to genuine simulation control, in the sense of producing r.v.'s approximately distributed from f or even i.i.d. from f. Both approaches have drawbacks, too, since the single chain methods can never guarantee[2] that the whole

---

[1] Thanks to Peter Green, we became aware of a subtle difference between *contrôle* (Fr.) and *control* (Eng.). We will thus use *control* in its French meaning of monitoring, evaluation or assessment.

[2] See, however, the Riemann control variate of Philippe, 1997a, described in §3.3, which escapes this difficulty by evaluating the probability of the region already explored by the Markov chain, based on a single path of this Markov chain.

support of f has been explored,[3] while parallel chain methods depend on the choice of the initial distribution and thus only get interesting for well-dispersed distributions. It is thus preferable, although more conservative, to advise for a cumulated implementation of these methods.

The last section (§2.4) introduces the notion of *coupling*, which is instrumental to perfect simulation techniques (see §1.4), and is also of potential value for convergence control, although we do not take advantage of it in this book.

2.2   Convergence assessments for single chains

2.2.1   Graphical evaluations

While a simple monitoring of the chain $(x^{(t)})$ can only expose strong non-stationarities, it is more relevant to consider the cumulated sums (1.2), since they need to stabilize for convergence to be achieved. While this is only a necessary condition for convergence, since a stabilization of the average (1.2) may only correspond to the exploration of a single mode of f by the chain, improved monitoring involves several estimates of (1.1) based on the same chain, as in Robert (1995). Possible alternatives to the empirical average (1.2) are *conditional* expectations *(Rao-Blackwellization)*, *importance sampling* and *quadrature methods* as in the *Riemann approximation technique* of Yakowitz *et al.* (1978) and Philippe (1997a), based on the order statistics $x_{(1)} \leq \ldots \leq x_{(T)}$ of $(x^{(1)}, \ldots, x^{(T)})$,

$$S_T^R = \sum_{t=1}^{T-1} [x_{(t+1)} - x_{(t)}] \, h(x_{(t)}) \, f(x_{(t)}), \qquad (2.1)$$

which is only relevant in dimension 1. Both alternatives are further discussed in Chapter 3, which also presents a control variate technique based on (2.1).

Example 2.2.1 Instead of $\pi(\theta) = 1$, consider a normal proper prior $\pi(\theta) =$
<small>CAUCHY</small> $\exp(-\theta^2/2\sigma^2)$ with known σ. The corresponding Gibbs sampler is associ-
<small>BENCHMARK</small>   ated with three artificial r.v.'s, η_1, η_2, η_3, such that

$$\pi(\theta, \eta_1, \eta_2, \eta_3 | x_1, x_2, x_3) \quad \propto \quad e^{-\theta^2/2\sigma^2} \, e^{-(1+(\theta-x_1)^2)\eta_1/2}$$
$$\times e^{-(1+(\theta-x_2)^2)\eta_2/2} \, e^{-(1+(\theta-x_3)^2)\eta_3/2} \; .$$

The conditional distributions are then

$$\eta_i | \theta, x_i \quad \sim \quad \mathcal{E}xp\left(\frac{1 + (\theta - x_i)^2}{2}\right), \qquad (i = 1, 2, 3)$$

---

[3] This is the *"You've only seen where you've been"* defect.

$$\theta | x_1, x_2, x_3, \eta_1, \eta_2, \eta_3 \quad \sim \quad \mathcal{N}\left(\frac{\eta_1 x_1 + \eta_2 x_2 + \eta_3 x_3}{\eta_1 + \eta_2 + \eta_3 + \sigma^{-2}}, \frac{1}{\eta_1 + \eta_2 + \eta_3 + \sigma^{-2}}\right).$$

We denote

$$\mu(\eta_1, \eta_2, \eta_3) = \frac{\eta_1 x_1 + \eta_2 x_2 + \eta_3 x_3}{\eta_1 + \eta_2 + \eta_3 + \sigma^{-2}}$$

and

$$\tau^{-2}(\eta_1, \eta_2, \eta_3) = \eta_1 + \eta_2 + \eta_3 + \sigma^{-2}$$

the conditional mean and variance of the conditional distribution of θ. When $h(\theta) = \exp(-\theta/\sigma)$, the different approximations of $\mathbb{E}_\pi[h(\theta)]$ are

$$S_T^C = \frac{1}{T} \sum_{t=1}^{T} \exp\left\{-\mu\left(\eta_1^{(t)}, \eta_2^{(t)}, \eta_3^{(t)}\right) + \tau^2\left(\eta_1^{(t)}, \eta_2^{(t)}, \eta_3^{(t)}\right)/2\right\},$$

for the conditional expectation, $S_T^I = \sum_{t=1}^{T} w_t h(\theta^{(t)}) / \sum_{t=1}^{T} w_t$, with

$$w_t \propto \frac{\exp\left\{-\dfrac{(\theta^{(t)})^2}{2\sigma^2} + \dfrac{\left(\theta^{(t)} - \mu\left(\eta_1^{(t)}, \eta_2^{(t)}, \eta_3^{(t)}\right)\right)^2}{2\tau^2\left(\eta_1^{(t)}, \eta_2^{(t)}, \eta_3^{(t)}\right)}\right\}}{\tau\left(\eta_1^{(t)}, \eta_2^{(t)}, \eta_3^{(t)}\right) \prod_{i=1}^{3}\left[1 + (x_i - \theta^{(t)})^2\right]},$$

for importance sampling and

$$S_T^R = \frac{\sum_{t=1}^{T-1} \left(\theta_{(t+1)} - \theta_{(t)}\right) e^{-\theta_{(t)}/\sigma - \theta_{(t)}^2/(2\sigma^2)} \prod_{i=1}^{3}\left[1 + (x_i - \theta_{(t)})^2\right]^{-1}}{\sum_{t=1}^{T-1} \left(\theta_{(t+1)} - \theta_{(t)}\right) e^{-\theta_{(t)}^2/(2\sigma^2)} \prod_{i=1}^{3}\left[1 + (x_i - \theta_{(t)})^2\right]^{-1}},$$

for the Riemann approximation, where $\theta_{(1)} \leq \ldots \leq \theta_{(T)}$ are the order statistics associated with the $\theta^{(t)}$'s.

Figure 2.1 describes the convergence of the four estimators as t increases. As often, S_T and S_T^C are quite similar from the start, S_T^R is very stable and shows that convergence is rapidly achieved for this example. The importance sampling estimate S_T^P has not yet converged, which can be explained by the infinite variance of the weights w_t. ‖

Yu and Mykland (1998) propose a graphical evaluation based on the *cumulated sums* (CUSUM's), which monitor the partial differences

$$D_T^i = \sum_{t=1}^{i} [h(x^{(t)}) - S_T], \qquad i = 1, \cdots, T,$$

to assess the mixing behavior of the chain and correlation between the $x^{(t)}$'s: the more mixing the chain is, the closer to Brownian motion the graph of

FIGURE 2.1. Convergence of the four estimators S_T (full), S_T^C (dots), S_T^R (dashes) and S_T^P (long dashes) for $\sigma^2 = 50$ and $(x_1, x_2, x_3) = (-8, 8, 17)$. The graphs for S_T and S_T^C are indistinguishable. The final values are 0.845, 0.844, 0.828 and 0.845 for S_T, S_T^C, S_T^P and S_T^R respectively. (*Source:* Robert, 1996c.)

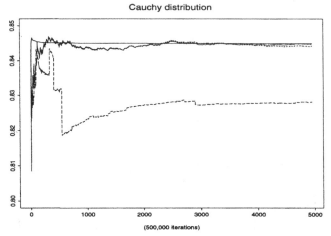

the D_T^i's is. For slowly mixing chains, the graphs are on the contrary much smoother, with long excursions away from 0. These very empirical rules are however too tentative for the method of Yu and Mykland (1998) to appear as a strong convergence criterion.

2.2.2   Binary approximation

Raftery and Lewis (1992a,b, 1996) have proposed a technique which pertains to our approach, namely to use some *finite Markov chain* theory to control the convergence of the chain of interest, by approximating the minimal time t_0 to reach convergence, the sample size T necessary to evaluate (1.1), as well as the "batch" size k, from the finite structure. (The *batch* size k gives the number of iterations ignored between two recordings of the Markov chain. This strategy is commonly used in dependent simulation to approximate independence. It is only validated for MCMC algorithms in the special case of *interleaving*, as shown by Liu, Wong and Kong, 1995.) The authors derive a two-state process from the Markov chain $(x^{(t)})$ as

$$z^{(t)} = \mathbb{I}_{x^{(t)} \leq \underline{x}} \,,$$

where $\underline{x}$ is an arbitrary point in the support of f. Assuming $(z^{(t)})$ is an homogeneous Markov chain (although this is not the case in general), with transition matrix

$$\mathbb{P} = \begin{pmatrix} 1 - \alpha & \alpha \\ \beta & 1 - \beta \end{pmatrix} \,,$$

the associated invariant distribution is

$$P(z^{(\infty)} = 0) = \frac{\beta}{\alpha + \beta} , \qquad P(z^{(\infty)} = 1) = \frac{\alpha}{\alpha + \beta} .$$

The warm-up time can then be deduced from the condition

$$\left| P(z^{(t_0)} = i | z^{(0)} = j) - P(z^{(\infty)} = i) \right| < \varepsilon$$

for $i, j = 0, 1$. Raftery and Lewis (1992a) show that this condition is equivalent to

$$|1 - \alpha - \beta|^{t_0} \leq \frac{(\alpha + \beta)\varepsilon}{\alpha \vee \beta} ,$$

i.e.

$$t_0 \geq \log \left(\frac{(\alpha + \beta)\varepsilon}{\alpha \vee \beta} \right) \bigg/ \log |1 - \alpha - \beta|.$$

For $h(z) = z$, the minimal sample size for the convergence of

$$\delta_T = \frac{1}{T} \sum_{t=t_0}^{t_0+T} z^{(t)}$$

to $\dfrac{\alpha}{\alpha + \beta}$ is derived from the normal approximation of δ_T, with variance

$$\frac{1}{T} \frac{(2 - \alpha - \beta) \, \alpha\beta}{(\alpha + \beta)^3} .$$

For instance,

$$P \left(\left| \delta_T - \frac{\alpha}{\alpha + \beta} \right| < q \right) \geq \varepsilon'$$

implies

$$\Phi \left(\sqrt{T} \frac{(\alpha + \beta)^{3/2} q}{\sqrt{\alpha\beta(2 - \alpha - \beta)}} \right) \geq \frac{\varepsilon' + 1}{2} ,$$

i.e.

$$T \geq \frac{\alpha\beta(2 - \alpha - \beta)}{q^2(\alpha + \beta)^3} \, \Phi^{-1} \left(\frac{\varepsilon' + 1}{2} \right) .$$

Since $(z^{(t)})$ is not a Markov chain, even when $x^{(t)}$ has a finite support (see Kemeny and Snell, 1960), Raftery and Lewis (1992a,b) determine a batch step k by testing whether $(z^{(kt)})$ is a Markov chain against the alternative hypothesis that $(z^{(kt)})$ is a second order Markov process. This derivation is rather weak from a theoretical point of view since the alternative hypothesis is restrictive and its rejection does not imply that the null hypothesis holds. Moreover, it is possible to construct a true binary Markov chain by subsampling at *renewal* times (see Chapter 4) rather than at fixed times. From a practical point of view, it appears in the examples (see Raftery and Lewis, 1992a, Brooks, 1998, Brooks and Roberts, 1997, and Cowles and Carlin, 1996) that the values of k obtained through these tests are usually quite small and often equal to 1.

The previous analysis and the derivation of the quantities (T_0, T, k) depends solely on the parameter (α, β), which is usually unknown. The binary control technique of Raftery and Lewis (1992a, b) thus requires a preliminary run where (α, β) is "correctly" estimated. While this is easier than for the original chain, since the state-space is reduced to two points, another control technique is necessary to decide whether (α, β) are indeed well-estimated. An alternative is to iteratively estimate (α, β) until all parameters stabilize (Raftery and Lewis, 1996). Comparing with the independent setup, Raftery and Lewis (1992a) suggest to use first a sample size larger than

$$ T_{\min} \geq \Phi^{-1} \left(\frac{\varepsilon' + 1}{2} \right)^2 \frac{\alpha\beta}{(\alpha + \beta)^2} q^{-1} . $$

A problem with this solution is that the Central Limit Theorem does not necessarily applies for T as small as this $T_{\min}$, as shown in Chapter 5.

As mentioned in Brooks and Roberts (1997), the method does not perform well when $\underline{x}$ is located in the queues of f. A sensitivity analysis must then be led to robustify the choice of $\underline{x}$ and this may considerably increase the computing time in large dimension models. Nonetheless, the binary control method presents the major incentive to be almost totally automated and it does not require additional programming time, given the programs already existing in `Statlib`.

TABLE 2.1. Parameters derived from the binary control for three possible parameterizations with control parameters $\epsilon = q = 0.005$ and $\epsilon' = 0.999$ (5000 preliminary runs).

parameter	$\underline{x}$	α	β	q_0	t_0	T
μ	0.13	0.19	0.34	0.36	6	42345
η	0.35	0.38	0.18	0.68	6	36689
ξ	0.02	0.29	0.40	0.42	4	30906

Example 2.2.2 For $a = (0.1, 0.14, 0.7, 0.9)$, $b = (0.17, 0.24, 0.19, 0.20)$ and $(x_1, x_2, x_4, x_5) = (4, 15, 12, 7, 4)$, Table 2.1 provides the values of α, β, $q_0 = \beta/(\alpha + \beta)$, t_0 and T for two-state variables derived from μ, η and $\xi = \mu\eta$, respectively. The preliminary run is of size 5000. The control parameters ϵ, ϵ' and q are rather strict and the corresponding values of T are large, with reduced initial sample sizes. In this case, the parameterization (η vs. μ vs. ξ) is negligible. ∥

MULTINOMIAL
BENCHMARK

2.3 Convergence assessments based on parallel chains

2.3.1 Introduction

As mentioned in §2.2, the single chain methods have the drawback that the chain hardly brings information on the regions of the space it does not visit. Parallel chains methods try to overcome this defect by generating in parallel M chains $(\theta_m^{(t)})$ $(1 \leq m \leq M)$, aiming at eliminating the dependence on initial conditions, and the convergence control is most often based on the comparison of the estimations of different quantities for the M chains, although Chapter 5 develops an alternative. An obvious danger of this approach is that the slowest chain commands the speed of convergence. The various methods proposed in the literature are actually far from free of defects (see Geyer, 1992), since a preliminary (partial) knowledge of the distribution of interest is crucial to ensure good performances. Indeed, an initial distribution which omits one or several important modes of f does not improve much over a single starting value if the MCMC algorithm under study has a strong tendency to get trapped near the starting mode. Another major drawback is that the comparison between two chains at time T is delicate, given that they are not converging at the same speed. For complex settings like the Gibbs sampler on nonlinear models, where the MCMC algorithm may be quite slow, it is thus more efficient to use a single chain of size MT rather than M chains of size T, which will likely remain in a neighborhood of their starting point. (See Tierney, 1994, and Raftery and Lewis, 1996, for additional criticisms.) The debate *parallel vs. single chain* is, however, far from being closed and most methods proposed in this monograph will rely on parallel chains, at one stage or another.

2.3.2 Between-within variance criterion

Gelman and Rubin (1992) initiate their control strategy by constructing an initial distribution μ related to the modes of f, obtained for instance by numerical methods. Their suggestion is to use a mixture of Student's t distributions centered at the modes of f and with scale parameters derived from the second derivatives of f at these modes. Using these as initial distributions, one generates M chains $(x_m^{(t)})$ $(1 \leq m \leq M)$. For every parameter of interest $\xi = h(x)$, Gelman and Rubin's (1992) criterion is based on the difference between a weighted estimator of the variance for each chain and the variance of the estimators on the different chains.

More precisely, consider

$$B_T = \frac{1}{M} \sum_{m=1}^{M} (\bar{\xi}_m - \bar{\xi})^2 ,$$

$$W_T \;=\; \frac{1}{M}\sum_{m=1}^{M} s_m^2 \;=\; \frac{1}{M}\sum_{m=1}^{M}\frac{1}{T}\sum_{t=1}^{T}(\xi_m^{(t)}-\bar\xi_m)^2 \;,$$

with

$$\bar\xi_m = \frac{1}{T}\sum_{t=1}^{T}\xi_m^{(t)}, \qquad \bar{\bar\xi} = \frac{1}{M}\sum_{m=1}^{M}\bar\xi_m$$

and $\xi_m^{(t)} = h(x_m^{(t)})$. The quantities B_T and W_T are the *between-* and *within-chain* variances. A first estimator of the posterior variance of ξ is

$$\hat\sigma_T^2 = \frac{T-1}{T}\,W_T + B_T \;.$$

Gelman and Rubin (1992) compare $\hat\sigma_T^2$ and W_T, which are asymptotically equivalent, through a Student approximation; $\hat\sigma_T^2$ overestimates the variance of the $\xi_m^{(t)}$'s because of the large dispersion of the initial distribution, while W_T underestimates this variance when the sequences $(\xi_m^{(t)})$ remain concentrated around their initial value. If we denote

$$
\begin{aligned}
\omega_T \;=\;& \left(\frac{T-1}{T}\right)^2 \frac{1}{M^2}\left[\sum_{m=1}^{M} s_m^4 - \frac{1}{M}\left(\sum_{m=1}^{M} s_m^2\right)^2\right] \\
&+ 2\,\frac{(M+1)(T-1)}{M^3 T}\left[\sum_{m=1}^{M} s_m^2\,\bar\xi_m^2 - \frac{1}{M}\sum_{m=1}^{M} s_m^2 \sum_{m=1}^{M}\bar\xi_m^2\right. \\
&\left. - 2\bar{\bar\xi}\sum_{m=1}^{M} s_m^2\,\bar\xi_m + 2\bar{\bar\xi}^2\sum_{m=1}^{M} s_m^2\right] + 2\,\frac{(M+1)^2}{M^2(M-1)}\,B_T^2 \;,
\end{aligned}
$$

with corresponding degree of freedom

$$\nu_T = 2\,\frac{\left(\hat\sigma_T^2 + \frac{B_T}{M}\right)^2}{\omega_T}$$

the criterion of Gelman and Rubin (1992) is given by

$$
\begin{aligned}
R_T \;=\;& \frac{\hat\sigma_T^2 + \frac{B_T}{M}}{W_T}\,\frac{\nu_T}{\nu_T - 2} \\
=\;& \left(\frac{T-1}{T} + \frac{M+1}{M}\,\frac{B_T}{W_T}\right)\frac{\nu_T}{\nu_T - 2}\;.
\end{aligned}
$$

Normal approximations lead to an approximate $\mathcal{F}(M-1,\psi_T)$ distribution for TB_T/W_T, with $\psi_T = 2W_T^2/\varpi_T$ and

$$\varpi_T = \frac{1}{M^2}\left[\sum_{m=1}^{M} s_m^4 - \frac{1}{M}\left(\sum_{m=1}^{M} s_m^2\right)^2\right].$$

A test of $\mathbb{E}[R_T] = 1$ can be derived from this approximation.

This method is commonly used, because of its simplicity and of its connections with standards tools of linear regression. Nonetheless, it suffers from several drawbacks. First, the detailed construction of μ is costly and unreliable, given that it usually implies advanced maximization techniques but cannot guarantee an exhaustive list of the modes of f. Second, and maybe more importantly, the evaluation of convergence is based on normal approximations, which are only valid asymptotically, are difficult to validate and are certainly inappropriate in some MCMC settings. Third, the criterion is even more difficult to derive and to assess in multidimensional problems, as shown by Brooks and Gelman (1998).

2.3.3   Distance to the stationary distribution

The following methods are rather "off-key", given our overall purpose, in the sense that they seek to estimate some distance to the stationary distribution, rather than focusing on the chain(s) at hand. For instance, in the case of the Gibbs sampler, Roberts (1992) proposes an unbiased estimator of the distance $\|f_t - f\|$, where f_t is the marginal density of the "symmetrized" chain $x^{(t)}$, which is obtained from a reversible version of the Gibbs sampler, namely by running the steps 1., 2.,$\cdots$,p. of $[A_2]$, then the steps p.,p-1.,$\cdots$,1.. This method associates to $(x^{(t)})$ a *dual* chain $(\tilde{x}^{(t)})$: given $x^{(t)}$, $\tilde{x}^{(t)}$ is simulated conditionally on $\theta^{(t)}$ through steps 1., 2.,$\cdots$,p. of $[A_2]$, then $x^{(t+1)}$ is simulated conditionally on $\tilde{x}^{(t)}$ through steps p.,p-1.,$\cdots$,1. of $[A_2]$.[4]

Starting with a single initial value $x^{(0)}$, Roberts' (1992) method is based on M parallel chains, $(x_\ell^{(t)})$ $(\ell = 1, \cdots, M)$, and on the following unbiased estimator of $\|f_t - f\| + 1$:

$$J_t = \frac{1}{M(M-1)} \sum_{1 \le \ell \neq s \le M} \frac{K_-(\tilde{x}_\ell^{(0)}, x_s^{(t)})}{f(x_s^{(t)})} \,,$$

where K_- denotes the transition kernel associated with steps p.,p-1.,$\cdots$, 1. of $[A_2]$. Since f is usually known up to a multiplicative constant, the limiting value of J_t is unknown (instead of being 1) and the control method is a simple graphical monitoring of the stabilization of J_t. Note that the normalizing constants of $K_-(x, x')$ have to be known since they depend on $\tilde{x}_\ell^{(0)}$, and this can be quite costly in terms of computing time.

This estimation method is theoretically well-grounded but, as mentioned above, it somehow misses the true purposes of convergence control, as the marginal distribution f_t is not of direct interest in the control of MCMC algorithms. The dependence on the *same* initial value is also a negative

---

[4]Note that these additional steps create a reversible chain.

feature of the method, because a slow mixing algorithm will give similar values of $x_s^{(t)}$ for small t's and thus leads to the impression that convergence is reached.

Roberts (1994) extends this method to other MCMC algorithms. Brooks, Dellaportas and Roberts (1997) propose a similar approach which estimates an upper bound on the distance L_1 between f_t and f, based on the following relation:

$$||f - f_t||_1 = \mathbb{E}_{f_t} \left[1 \wedge \frac{f(x)}{f_t(x)} \right]$$

(see also Brooks and Roberts, 1998).

As in Roberts (1992), Liu, Liu and Rubin (1992) evaluate the difference between f and f_t. Their method is based on an unbiased estimator of the variance of $f_t(\theta)/f(\theta)$, namely $U - 1$, with

$$U = \frac{f(\theta_1)}{f(\theta_2)} \frac{K(\theta_1^-, \theta_2)}{K(\theta_1^-, \theta_1)} ,$$

where $\theta_1^-, \theta_2^- \sim f_{(t-1)}$ are independent, $\theta_1 \sim K(\theta_1^-, \theta_1)$ and $\theta_2 \sim K(\theta_2^-, \theta_2)$. Using M parallel chains, each iteration t provides $M(M-1)$ values $U^{(i,j,t)}$ $(i \neq j)$ which can be monitored graphically or with Gelman and Rubin's (1992) method. Note that the ratio U does not imply the computation of the normalizing constant for the kernel K.

The following section (§2.4) provides another approach to the estimation of the total variation distance which has not been yet exploited in MCMC settings.

2.4   Coupling techniques

Although the notion of *coupling* is not used as a control technique *per se* in this book, it appears at several places (Chapters 4 and 6). This probabilistic notion is indeed related to convergence control in the sense that it establishes independence from initial conditions and may as well accelerate convergence. Besides, it is related with the exact sampling techniques of §1.4. We thus give a short introduction on coupling at this stage, before presenting the coupling control method of Johnson (1996). Its potential for actual convergence control should not be neglected, even if we only marginally use coupling in this monograph.

2.4.1   Coupling theory

While the general purpose of *coupling* is to evaluate the distance between two distributions by creating a joint distribution whose marginals are the distributions of interest (see Lindvall, 1992), its relevance in our setting is both to evaluate the (total variation) distance to the stationary distribution and to accelerate convergence to this distribution.

Definition 2.4.1 Two Markov chains $(x_1^{(t)})$ and $(x_2^{(t)})$ with initial distributions μ_1 and μ_2 are *coupled* if the joint distribution of $(x_1^{(t)}, x_2^{(t)})$ preserves the marginal distributions of $(x_1^{(t)})$ and $(x_2^{(t)})$. A *coupling time* is thus a stopping time T such that

$$x_1^{(t)} = x_2^{(t)} \qquad \text{for} \quad t \geq T .$$

Obviously, a coupling time is only useful if it is almost surely finite. (The coupling is then said to be *successful*.) Therefore, this notion seems to solely apply in specific settings like finite or discrete chains. However, if the transition kernel is such that there exists a *recurrent atom* α, i.e. an accessible set such that $q(\cdot|x)$ is constant for $x \in \alpha$, both chains $(x_1^{(t)})$ and $(x_2^{(t)})$ can be made identical once they meet in α. (A weaker notion of coupling requires that the distributions of $x_1^{(t)}$ and $x_2^{(t)}$ are identical for $t \geq T$.) While atoms are rarely encountered in (continuous) MCMC setups, Chapter 4 shows that the existence of *small sets* is sufficient to achieve the same goal, namely that two independent chains meet with positive probability at a renewal time (see §4.2.1). More generally, we will see below that a *maximal coupling* algorithm provides a generic way to couple two Markov chains.

As shown by Lindvall (1992), the distribution of the coupling time is strongly related to the total variation distance between the distributions of $(x_1^{(t)})$ and $(x_2^{(t)})$, in the sense of *the fundamental coupling inequality*,

$$\|P_{\mu_1}^n - P_{\mu_2}^n\|_{TV} \leq 2P(T > n), \qquad (2.2)$$

where $P_{\mu_i}^n$ denotes the distribution of $x_i^{(n)}$ $(i = 1, 2)$. If μ_2 is the stationary distribution f, the inequality (2.2) is

$$\|P_{\mu_1}^n - f\|_{TV} \leq 2P(T > n).$$

An empirical study of the distribution of T thus provides information on the convergence of $P_{\mu_1}^n$ to the stationary distribution.

Two particular types of couplings are *Doeblin's coupling*, where two independent chains are monitored till they meet (in an arbitrary state, in an atom, or in a small set with a certain renewal probability), and the *deterministic coupling* where $x_2^{(t)}$ is a deterministic function of $x_1^{(t)}$ (conditionally on $x_2^{(t-1)}$) through the use of the same underlying uniform variable.[5] For

---

[5] This obviously excludes the use of standard simulation techniques like accept-reject methods (see Robert and Casella, 1998).

instance, if the transition kernel corresponds to an exponential distribution

$$y|x \sim \mathcal{E}xp(x/(1+x)),$$

$x_1^{(t)}$ can be constructed as

$$x_1^{(t)} = -\frac{1 + x_1^{(t-1)}}{x_1^{(t-1)}} \log(u_t),$$

where $u_t \sim \mathcal{U}_{[0,1]}$, and $x_2^{(t)}$ is then

$$
\begin{aligned}
x_2^{(t)} &= -\frac{1 + x_2^{(t-1)}}{x_2^{(t-1)}} \log(u_t) \\
&= \frac{1 + x_2^{(t-1)}}{x_2^{(t-1)}} \frac{x_1^{(t-1)} x_1^{(t)}}{1 + x_1^{(t-1)}}
\end{aligned}
$$

(Both chains are thus identical once they have met.)

Another type of coupling has been developed for probabilistic reasons. It is called *maximal coupling* because it leads to an equality[6] in (2.2). The algorithm corresponding to the maximal coupling is as follows:

1. **Generate** $x_1^{(t)} \sim q(x|x_1^{(t-1)})$ **and** $u_t \sim \mathcal{U}_{[0,1]}$.

2. **If** $u_t \, q(x_1^{(t)}|x_1^{(t-1)}) \leq q(x_1^{(t)}|x_2^{(t-1)})$, **take** $x_2^{(t)} = x_1^{(t)}$. $[A_6]$

3. **Else, generate** $x_2^{(t)} \sim q(x|x_2^{(t-1)})$ **and** $u_t' \sim \mathcal{U}_{[0,1]}$
   until $u_t' q(x_2^{(t)}|x_2^{(t-1)}) \geq q(x_2^{(t)}|x_1^{(t-1)})$.

It is relevant for our purposes to establish that the algorithm $[A_6]$ is truly a coupling algorithm, namely that

$$x_1^{(t)} \sim q(x|x_1^{(t-1)}) \qquad \text{and} \qquad x_2^{(t)} \sim q(x|x_2^{(t-1)})$$

because the proof brings new lights on some results of Chapter 4 (§4.2.3 and Lemma 4.2.2). In fact, while $x_1^{(t)} \sim q(x|x_1^{(t-1)})$ by construction, the density of $x_2^{(t)}$ is indeed

$$
\begin{aligned}
q(x|x_1^{(t-1)}) \; \wedge \; & q(x|x_2^{(t-1)}) + (1 - \epsilon)q(x|x_2^{(t-1)}) \\
& \times \sum_{i=0}^{\infty} \epsilon^i \left(1 - \frac{q(x|x_1^{(t-1)}) \wedge q(x|x_2^{(t-1)})}{q(x|x_2^{(t-1)})} \right) \\
= \; & q(x|x_2^{(t-1)}),
\end{aligned}
$$

---

[6]Note that this does not signify that coupling occurs faster or that it is optimal in any sense. The appeal of this particular coupling is to provide a more accurate connection between total variation distance and coupling time.

where ϵ is the normalizing constant for $q(x|x_1^{(t-1)}) \wedge q(x|x_2^{(t-1)})$, i.e.

$$\epsilon = \int q(x|x_1^{(t-1)}) \wedge q(x|x_2^{(t-1)}) dx \quad (\leq 1).$$

As in the splitting process of §4.2.3, the density $q(x|x_2^{(t-1)})$ is thus split into two parts,

$$q(x|x_2^{(t-1)}) = \epsilon \frac{q(x|x_1^{(t-1)}) \wedge q(x|x_2^{(t-1)})}{\epsilon}$$
$$+(1-\epsilon)\frac{q(x|x_2^{(t-1)}) - q(x|x_1^{(t-1)}) \wedge q(x|x_2^{(t-1)})}{1-\epsilon}$$

and the generation does not require ϵ to be known, which is quite appealing in simulation setups. Note however that both transitions must be known up to the *same* multiplicative constant. Moreover, in the event ϵ is close to 1, this scheme may require extremely long runs and alternatives based on $q(x|x_2^{(t-1)}) - q(x|x_1^{(t-1)}) \wedge q(x|x_2^{(t-1)})$ may be preferable, when both conditional distributions $q(x|x_2^{(t-1)})$ and $q(x|x_1^{(t-1)})$ are available.

From the point of view of MCMC algorithms, and in particular for the Gibbs sampler, coupling can be implemented in many ways. First, for a given order on the g_i's in $[A_2]$, the chain of the full vector y_1 can be coupled with another chain y_2 as in Examples 2.4.1–2.4.3 below. The basic coupling ratio is then

$$\frac{g_1(y_{11}^{(t)}|y_{22}^{(t-1)},\ldots,y_{2p}^{(t-1)})g_2(y_{12}^{(t)}|y_{11}^{(t)},y_{23}^{(t-1)},\ldots,y_{2p}^{(t-1)})}{g_1(y_{11}^{(t)}|y_{12}^{(t-1)},\ldots,y_{1p}^{(t-1)})g_2(y_{12}^{(t)}|y_{11}^{(t)},y_{23}^{(t-1)},\ldots,y_1^{(t-1)})}$$
$$\times \ldots \frac{g_p(y_{1p}^{(t)}|y_{11}^{(t)},\ldots,y_{1(p-1)}^{(t)})}{g_p(y_{1p}^{(t)}|y_{11}^{(t)},\ldots,y_{1(p-1)}^{(t)})}.$$

An alternative approach is to couple both chains at each stage j. $(1 \leq j \leq p)$ of the Gibbs sampler in $[A_2]$. The component $y_{2j}^{(t)}$ is then equal to $y_{1j}^{(t)}$ with probability

$$\frac{g_j(y_{1j}^{(t)}|y_{21}^{(t)},\ldots,y_{2(j-1)}^{(t)},y_{2(j+1)}^{(t-1)},\ldots,y_{2p}^{(t-1)})}{g_j(y_{1j}^{(t)}|y_{11}^{(t)},\ldots,y_{1(j-1)}^{(t)},y_{1(j+1)}^{(t-1)},\ldots,y_{1p}^{(t-1)})} \wedge 1 \, ,$$

the previous components of $y_2^{(t)}$ being independently equal to those of $y_1^{(t)}$.

Example 2.4.1 The implementation of the maximal coupling algorithm Cauchy Benchmark implies running two parallel chains $(\eta_1^{(t)}, \theta_1^{(t)})$ and $(\eta_2^{(t)}, \theta_2^{(t)})$ such that the former is run from the conditional distributions as in Example 1.2.3 and the later is generated from

1. Take

$$(\eta_2^{(t)}, \theta_2^{(t)}) = (\eta_1^{(t)}, \theta_1^{(t)})$$

with probability

$$\frac{\pi(\eta_1^{(t)}, \theta_1^{(t)} | \eta_2^{(t-1)}, \theta_2^{(t-1)})}{\pi(\eta_1^{(t)}, \theta_1^{(t)} | \eta_1^{(t-1)}, \theta_1^{(t-1)})} = \frac{\pi(\eta_1^{(t)} | \theta_2^{(t-1)})}{\pi(\eta_1^{(t)} | \theta_1^{(t-1)})}$$

$$= \prod_{i=1}^{3} \frac{1 + (\theta_2^{(t-1)} - x_i)^2}{1 + (\theta_1^{(t-1)} - x_i)^2} \exp\left\{ -\frac{1}{2}(\theta_2^{(t-1)2} \right.$$

$$\left. -\theta_1^{(t-1)2}) \sum_i \eta_{1i}^{(t)} + (\theta_2^{(t-1)} - \theta_1^{(t-1)}) \sum_i \eta_{1i}^{(t)} x_i \right\}$$

2. Else, generate

$$(\eta_2^{(t)}, \theta_2^{(t)}) \sim \pi(\eta_2^{(t)}, \theta_2^{(t)} | \eta_2^{(t-1)}, \theta_2^{(t-1)}), \qquad u \sim \mathcal{U}_{[0,1]}$$

until

$$u \geq \prod_{i=1}^{3} \frac{1 + (\theta_1^{(t-1)} - x_i)^2}{1 + (\theta_2^{(t-1)} - x_i)^2}$$

$$\times \exp\left\{ -\frac{1}{2}(\theta_1^{(t-1)2} - \theta_2^{(t-1)2}) \sum_i \eta_{2i}^{(t)} + (\theta_1^{(t-1)} - \theta_2^{(t-1)}) \sum_i \eta_{2i}^{(t)} x_i \right\}.$$

In order to study the speed of convergence to stationarity, we run the chain $(\eta_1^{(t)}, \theta_1^{(t)})$ with no modification at coupling times and the second chain $(\eta_2^{(t)}, \theta_2^{(t)})$ with restarts from the initial distribution $\mathcal{U}_{[x_{(1)}, x_{(3)}]}$ at every coupling time. We can then evaluate the coupling time as well as the total variation distance between $\mathcal{U}_{[x_{(1)}, x_{(3)}]}$ and the stationary distribution (under maximal coupling). When implemented with the same observations x_i as in Example 1.2.3, the algorithm leads to an average coupling time of 9.32 iterations (for 25,000 replications). Note that the distribution of the chain at the time of coupling is *not* the stationary distribution, as shown by Figure 2.2(a), which gives a sample of 50,000 values at coupling, along with the true stationary distribution. Although a given $\theta_1^{(t)}$ is approximately distributed from the stationary distribution, the $\theta_1^{(t)}$'s at coupling times are not, because of the bias created by the stopping rule.

The same approach can be implemented for the reverse order Gibbs sampling, namely on the chain $(\theta_1^{(t)}, \eta_1^{(t)})$ and $(\theta_2^{(t)}, \eta_2^{(t)})$. The coupling steps are then

1. Take

$$(\theta_2^{(t)}, \eta_2^{(t)}) = (\theta_1^{(t)}, \eta_1^{(t)})$$

with probability

$$\frac{\pi(\theta_1^{(t)}|\eta_2^{(t-1)})}{\pi(\theta_1^{(t)}|\eta_1^{(t-1)})} = \frac{\sigma(\eta_1^{(t-1)})}{\sigma(\eta_2^{(t-1)})} \frac{\exp\left(\theta_1^{(t)} - \mu(\eta_1^{(t-1)})\right)^2 / 2\sigma(\eta_1^{(t-1)})^2}{\exp\left(\theta_1^{(t)} - \mu(\eta_2^{(t-1)})\right)^2 / 2\sigma(\eta_2^{(t-1)})^2}$$

where

$$\sigma(\eta)^2 = \frac{1}{\eta_1 + \eta_2 + \eta_3 + \tau^{-2}}, \qquad \mu(\eta) = \frac{\eta_1 x_1 + \eta_2 x_2 + \eta_3 x_3}{\eta_1 + \eta_2 + \eta_3 + \tau^{-2}}.$$

2. Else, generate

$$(\theta_2^{(t)}, \eta_2^{(t)}) \sim \pi(\theta_2^{(t)}, \eta_2^{(t)}|\eta_2^{(t-1)}), \qquad u \sim \mathcal{U}_{[0,1]}$$

until

$$u > \frac{\sigma(\eta_2^{(t-1)})}{\sigma(\eta_1^{(t-1)})} \frac{\exp\left(\theta_2^{(t)} - \mu(\eta_2^{(t-1)})\right)^2 / 2\sigma(\eta_2^{(t-1)})^2}{\exp\left(\theta_2^{(t)} - \mu(\eta_1^{(t-1)})\right)^2 / 2\sigma(\eta_1^{(t-1)})^2}.$$

In this case, the average coupling time 9.66 is approximately the same, using the same uniform starting distribution on $\theta_2^{(0)}$'s. As shown by Figure 2.2(b), the distribution of the points at coupling is again different from the stationary distribution, although the difference is not as important as for the direct order.

FIGURE 2.2. Histogram of a sample of θ's of size $25,000$ at coupling time, obtained without restarts of the first chain and with a uniform $\mathcal{U}_{[x_{(1)}, x_{(3)}]}$ initial distribution, against the stationary distribution, (a) for the (η, θ) order and (b) for the (θ, η) order.

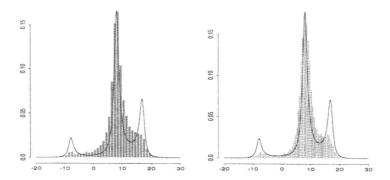

As mentioned above, coupling can be implemented at each stage of the Gibbs sampler, with the same acceptance probabilities as above. The improvement brought by this "double" coupling is rather marginal since the mean coupling time is then 8.39. This shows however that double coupling does not necessarily have a negative impact on coupling times. ‖

Example 2.4.2 A second chain $(\lambda_2^{(t)}, \beta_2^{(t)})$, with $\lambda_i = (\lambda_{i1}, \ldots, \lambda_{i10})$, can be coupled to the original chain $(\lambda_1^{(t)}, \beta_1^{(t)})$ as follows:

1. Take

$$(\lambda_2^{(t)}, \beta_2^{(t)}) = (\lambda_1^{(t)}, \beta_1^{(t)})$$

with probability

$$
\frac{\pi(\lambda_1^{(t)}, \beta_1^{(t)} | \lambda_2^{(t-1)}, \beta_2^{(t-1)})}{\pi(\lambda_1^{(t)}, \beta_1^{(t)} | \lambda_1^{(t-1)}, \beta_1^{(t-1)})} = \frac{\pi(\beta_1^{(t)} | \lambda_2^{(t-1)})}{\pi(\beta_1^{(t)} | \lambda_1^{(t-1)})}
$$

$$
= \left(\frac{\delta + \sum_i \lambda_{2i}^{(t-1)}}{\delta + \sum_i \lambda_{1i}^{(t-1)}} \right)^{\gamma + 10\alpha} e^{\beta_1^{(t)} \left(\sum_i \lambda_{1i}^{(t-1)} - \sum_i \lambda_{2i}^{(t-1)} \right)}
$$

2. Else, generate

$$(\lambda_2^{(t)}, \beta_2^{(t)}) \sim \pi(\lambda_2^{(t)}, \beta_2^{(t)} | \lambda_2^{(t-1)}), \qquad u \sim \mathcal{U}_{[0,1]}$$

until

$$
u > \left(\frac{\delta + \sum_i \lambda_{1i}^{(t-1)}}{\delta + \sum_i \lambda_{2i}^{(t-1)}} \right)^{\gamma + 10\alpha} e^{\beta_2^{(t)} \left(\sum_i \lambda_{2i}^{(t-1)} - \sum_i \lambda_{1i}^{(t-1)} \right)}
$$

For the dataset given in Table 1.2, and the uniform distribution on $[1.5, 3.5]$ as initial distribution, the average coupling time for $10,000$ replications is 1.61.

The reverse coupling is given by

1. Take

$$(\beta_2^{(t)}, \lambda_2^{(t)}) = (\beta_1^{(t)}, \lambda_1^{(t)})$$

with probability

$$
\frac{\pi(\lambda_1^{(t)} | \beta_2^{(t-1)})}{\pi(\lambda_1^{(t)} | \beta_1^{(t-1)})} = \prod_{i=1}^{10} \left(\frac{t_i + \beta_2^{(t-1)}}{t_i + \beta_1^{(t-1)}} \right)^{p_i + \alpha} e^{\lambda_{1i}^{(t)} (\beta_1^{(t-1)} - \beta_2^{(t-1)})}
$$

2. Else, generate

$$(\beta_2^{(t)}, \lambda_2^{(t)}) \sim \pi(\beta_2^{(t)}, \lambda_2^{(t)}), \qquad u \sim \mathcal{U}_{[0,1]}$$

until

$$
u > \prod_{i=1}^{10} \left(\frac{t_i + \beta_1^{(t-1)}}{t_i + \beta_2^{(t-1)}} \right)^{p_i + \alpha} e^{\lambda_{2i}^{(t)} (\beta_2^{(t-1)} - \beta_1^{(t-1)})}
$$

The implementation of this scheme leads to an average coupling time of 1.8, only slightly slower than the reverse approach, "despite" the larger dimension of the parameter λ. Double-coupling reduces the average coupling time to 1.47.   ‖

Example 2.4.3 If we denote $\theta = (\mu, \eta)$ and $z_1 = (z_{11}, z_{12}, z_{13}, z_{14})$, the chain $(\theta_1^{(t)}, z_1^{(t)})$ can be coupled with a second chain $(\theta_2^{(t)}, z_2^{(t)})$ as follows: MULTINOMIAL BENCHMARK

1. Take $(\theta_2^{(t)}, z_2^{(t)}) = (\theta_1^{(t)}, z_1^{(t)})$ with probability

$$\frac{\pi(\theta_1^{(t)} | z_2^{(t-1)})}{\pi(\theta_1^{(t)} | z_1^{(t-1)})} = \frac{\Gamma(z_{\mu 2}^{(t-1)} + z_{\eta 2}^{(t-1)} + x_5 + 1.5)}{\Gamma(z_{\mu 1}^{(t-1)} + z_{\eta 1}^{(t-1)} + x_5 + 1.5)}$$

$$\times \frac{\Gamma(z_{\mu 1}^{(t-1)} + 0.5)\Gamma(z_{\eta 1}^{(t-1)} + 0.5)}{\Gamma(z_{\mu 2}^{(t-1)} + 0.5)\Gamma(z_{\eta 2}^{(t-1)} + 0.5)}$$

$$\times \left(\mu_1^{(t)}\right)^{z_{\mu 2}^{(t-1)} - z_{\mu 1}^{(t-1)}} \left(\eta_1^{(t)}\right)^{z_{\eta 2}^{(t-1)} - z_{\eta 1}^{(t-1)}}$$

where $(i = 1, 2)$

$$z_{\mu i}^{(t-1)} = z_{i1}^{(t-1)} + z_{i2}^{(t-1)}, \qquad z_{\eta i}^{(t-1)} = z_{i3}^{(t-1)} + z_{i4}^{(t-1)}.$$

2. Else, generate

$$(\theta_2^{(t)}, z_2^{(t)}) \sim \pi(\theta, z | z_2^{(t-1)})$$

until

$$u > \frac{\Gamma(z_{\mu 1}^{(t-1)} + z_{\eta 1}^{(t-1)} + x_5 + 1.5)}{\Gamma(z_{\mu 2}^{(t-1)} + z_{\eta 2}^{(t-1)} + x_5 + 1.5)} \frac{\Gamma(z_{\mu 2}^{(t-1)} + 0.5)\Gamma(z_{\eta 2}^{(t-1)} + 0.5)}{\Gamma(z_{\mu 1}^{(t-1)} + 0.5)\Gamma(z_{\eta 1}^{(t-1)} + 0.5)}$$

$$\times \left(\mu_2^{(t)}\right)^{z_{\mu 1}^{(t-1)} - z_{\mu 2}^{(t-1)}} \left(\eta_2^{(t)}\right)^{z_{\eta 1}^{(t-1)} - z_{\eta 2}^{(t-1)}}$$

When implemented on the original dataset of Example 1.3.5, the average coupling time on $25,000$ iterations with a Dirichlet $\mathcal{D}(1, 1, 1)$ distribution on $\theta_2^{(0)}$ is 2.67. Note that the implementation of the coupling method requires a tight control of the normalizing constants which may be too computationally demanding in some cases.

The reverse order coupling is

1. Take $(z_2^{(t)}, \theta_2^{(t)}) = (z_1^{(t)}, \theta_1^{(t)})$ with probability

$$\frac{\pi(z_1^{(t)} | \theta_2^{(t-1)})}{\pi(z_1^{(t)} | \theta_1^{(t-1)})} = \prod_{i=1,2} \left(\frac{\mu_2}{\mu_1}\right)^{z_{1i}^{(t)}} \left(\frac{a_i \mu_1 + b_i}{a_i \mu_2 + b_i}\right)^{x_i} \prod_{i=3,4} \left(\frac{\eta_2}{\eta_1}\right)^{z_{1i}^{(t)}} \left(\frac{a_i \eta_1 + b_i}{a_i \eta_2 + b_i}\right)^{x_i}$$

2. Else, generate

$$(z_2^{(t)}, \theta_2^{(t)}) \sim \pi(z, \theta | \theta_2^{(t-1)})$$

until

$$u > \prod_{i=1,2} \left(\frac{\mu_1}{\mu_2}\right)^{z_{2i}^{(t)}} \left(\frac{a_i\mu_2 + b_i}{a_i\mu_1 + b_i}\right)^{x_i} \prod_{i=3,4} \left(\frac{\eta_1}{\eta_2}\right)^{z_{2i}^{(t)}} \left(\frac{a_i\eta_2 + b_i}{a_i\eta_1 + b_i}\right)^{x_i}$$

In this order, the average coupling time is quite similar since it is equal to 2.71. To reproduce the comparison of Example 2.4.1, we integrated the posterior in η and μ to obtain the respective marginals in μ and η, using **Mathematica** for the case $\alpha_1 = \ldots = \alpha_3 = 1$ (see also Robert, 1995). The comparison between the true posteriors and the histograms of the $\theta^{(t)}$'s at coupling are more satisfactory than in Example 2.4.1, although there are still discrepancies around 0 and 1, as shown by Figure 2.3. A side consequence of this study is to show that the posterior marginals in μ and η are almost identical analytically. Double-coupling is much faster in this case since the average coupling time is then 1.95.   ∥

FIGURE 2.3. Histogram of a sample of μ's *(left)* and η's *(right)* of size 25,000, at coupling time, obtained with a $\mathcal{D}(1,1,1)$ initial distribution, against the stationary distribution, *(top)* for the (θ, z) order and *(middle)* for the (z, θ) order. For comparison purposes, the histograms of the whole samples of μ's and η's are also plotted against the stationary distribution *(bottom)*.

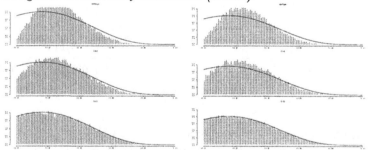

2.4.2   Coupling diagnoses

Although there are strong misgivings about the pertinence of a coupling strategy as a mean to start "in" the stationary distribution,[7] coupling and in particular optimal coupling can be used to evaluate the warmup time to stationarity. As shown by Example 2.4.1, if one of the two chains is run without interruption, that is with no restart at each coupling event, the average coupling time gives a convergent estimator of the mean number of iterations till stationarity, while the evaluation of the total variation

---

[7]Unless one uses backward coupling as in perfect simulation (see §1.4).

distance $||\mu P^n - \pi||_{TV}$ is more challenging, as seen in Johnson's (1996) attempts (see below). Note that different coupling strategies can be compared by this device, although independent coupling cannot work in continuous setups since the probability of coupling is then 0.

An additional use of coupling is to compare different orderings in Gibbs setups. More precisely, given a decomposition $(y_1, \ldots, y_p)$ of y and the corresponding conditional distributions $g_1, \ldots, g_p$ of f, as in $[A_2]$, there are $p!$ ways of implementing the Gibbs sampler, by selecting the order in which the components are generated. It is well-known that this order is not innocuous, and that some strategies are superior to others. The superiority of random scan Gibbs samplers, where the successive components are chosen at random, either independently or in a multinomial fashion–which amounts to select a random permutation–, has also been shown by Liu, Roberts and others. Since the practical consequences of these results are poorly known, the implementation of the corresponding sampling schemes has been rather moderate. Now, for a given ordering (or a given distribution for a random scan), the coupling time can be assessed by the above method, and this produces a practical comparison of various scans which should be useful in deciding which scheme to adopt. That important differences may occur has already been demonstrated in Examples 2.4.1–2.4.3.

Johnson (1996) suggests to use coupling based on M parallel chains $(\theta_m^{(t)})$ $(1 \leq m \leq M)$ which are functions *of the same* sequence of uniform variables. (This is a particular case of the *deterministic coupling* mentioned above.) For the Gibbs sampler $[A_2]$, the coupling method can thus be written as

1. **Generate M initial values $\theta_m^{(0)}$ $(1 \leq m \leq M)$.**

2. **For $1 \leq i \leq p$, $1 \leq m \leq M$, generate $\theta_{i,m}^{(t)}$ from** $[A_7]$

$$\theta_{i,m}^{(t)} = F_i^{-1}(u_i^{(t)}|\theta_{1,m}^{(t)}, \cdots, \theta_{i-1,m}^{(t)}, \theta_{i+1,m}^{(t-1)}, \cdots, \theta_{p,m}^{(t-1)}) .$$

3. **Stop the iterations when**

$$\theta_1^{(T)} = \cdots = \theta_M^{(T)} .$$

In a continuous setup, the stopping rule **3.** must be replaced by the approximation

$$\max_{m,n} |\theta_m^{(T)} - \theta_n^{(T)}| < \epsilon,$$

for $1 \leq m, n \leq M$ (or by the simultaneous visit to an atom, or yet by the simultaneous occurrence of a renewal event). In the algorithm $[A_7]$, F_i denotes the cdf of the conditional distribution $g_i(\theta_i|\theta_1, \cdots, \theta_{i-1}, \theta_{i+1}, \cdots, \theta_p)$. A necessary condition for Johnson's (1996) method to apply is thus that the conditional distributions must be simulated by *inversion* of the cdf, which is a rare occurrence.[8]

As shown in Examples 2.4.1–2.4.3, this method induces in addition strong biases, besides being strongly dependent on the initial distribution.

---

[8] For instance, this excludes Metropolis–Hastings algorithms, a drawback noticed by Johnson (1996), as well as all accept-reject algorithms.

3
Linking Discrete and Continuous Chains

Anne Philippe
Christian P. Robert

3.1 Introduction

When comparing discrete and continuous Markov chains from a theoretical perspective (through, say, Kemeny and Snell, 1960, or Feller, 1970, vol. 1, for the former and Revuz, 1984, or Meyn and Tweedie, 1993, for the latter), a striking difference is the scale of the machinery needed to deal with continuous Markov chains and, as a corollary, the relative lack of intuitive basis behind theoretical results for continuous Markov chains. This gap is the major incentive for this book, in the sense that convergence controls methods must keep away both from the traps of *ad hoc* devices which are "seen" to work well on artificial and contrived examples, and from the quagmire of formal convergence results which, while being fascinating from a theoretical point of view, either fail to answer the true purpose of the analysis, i.e. to decide whether or not the chain(s) at hand have really converged, or require such an involved analysis that they are not customarily applicable besides case-study setups. This is also why techniques such as Raftery and Lewis (1996) are quite alluring, given their intuitive background and theoretical (quasi-)validity.

Chapter 4 will develop a general technique to produce finite state space Markov chains derived from (almost) arbitrary Markov chains. In the meanwhile, this chapter introduces the main theme of the book, namely the idea to rely on finite (or discrete) Markov chains to control convergence for general MCMC Markov chains. This is also related to the previous development of §1.5 on the Duality Principle. As mentioned in this section, there exists a particular class of MCMC algorithms where a discrete Markov chain is, naturally or artificially, part of the process generated by the MCMC algorithm and where its study provides sufficient information on the whole process. Most of these cases stem from the Data Augmentation algorithm introduced by Tanner and Wong (1987) and exposed in Tanner (1996). Recall that Data Augmentation is defined as the special version of the Gibbs sampler $[A_2]$ where the conditional distributions are only 2, so that the Markov chain is of the form[1] $(z^{(t)}, \theta^{(t)})$, with $z^{(t)} \sim f(z|\theta^{(t-1)})$ and $\theta^{(t)} \sim \pi(\theta|z^{(t)})$.

3.2   Rao-Blackwellization

The Duality Principle of §1.5, which checks convergence properties via the simplest chain, can also be explained through Rao-Blackwellization, introduced in §1.3. In fact, as suggested by Gelfand and Smith (1990) and mentioned in §1.3, it is sometimes preferable to consider the sums of conditional expectations

$$\frac{1}{T} \sum_{t=1}^{T} \mathbb{E}^{\pi^t}[h(\theta)|z^{(t)}] = \frac{1}{T} \sum_{t=1}^{T} \tilde{h}(z^{(t)}), \qquad (3.1)$$

rather than the direct averages $\sum_{t=1}^{T} h(\theta^{(t)})/T$, since the integration leading to (3.1) reduces the variance of the estimate, while keeping the unbiasedness. As detailed below, Liu, Wong and Kong (1994) give some sufficient conditions for this improvement to hold for every convex loss function. Therefore, when Rao-Blackwellization is justified theoretically and when $\tilde{h}$ can be written explicitly—a far-from-trivial requirement—, the convergence of (3.1) to the expected value $\mathbb{E}^{\pi}[h(\theta)]$ is indeed directed by the convergence properties of $(z^{(t)})$, not those of $(\theta^{(t)})$. In particular, it can be directly controlled by the Central Limit Theorem when $(z^{(t)})$ is finite (see also Chapter 5). As mentioned in §1.3, (3.1) can always be studied in parallel with the empirical average, whichever estimator has a smaller variance, since it provides a *control variate* (in a weak sense), given that both estimates converge to the same value. It is, however, often the case that they cannot be distinguished (see, e.g., Example 2.2.1).

Note that, when Rao-Blackwellization does not apply, usual averages can still be directed by the chain $(z^{(t)})$, as shown by the following result:

Theorem 3.2.1 *If $(z^{(t)})$ is geometrically convergent with compact state space and rate ϱ, for every $h \in L_2(\pi)$, there exists C_h such that*

$$|| \mathbb{E}^{\pi^t}[h(\theta)] - \mathbb{E}^{\pi}[h(\theta)] ||_2 < C_h \varrho^t.$$

Proof. Without loss of generality, consider the case when h is a real-valued

---

[1] The change in the notations, when compared with the previous chapters, is intended to keep up with latent variable model conventions, where z stands for the missing variable vector and θ for the parameter to be estimated (through simulation). This is also why we will denote the stationary distribution of θ by π rather than f.

function. Then

$$\left(\mathbb{E}^{\pi^t}[h(\theta)] - \mathbb{E}^{\pi}[h(\theta)]\right)^2 = \left(\int h(\theta)(\pi^t(\theta) - \pi(\theta))d\theta\right)^2$$

$$= \left(\int \int h(\theta)\pi(\theta|z)d\theta(f^t(z) - f(z))dz\right)^2$$

$$\leq \max_z \left(\mathbb{E}^{\pi}[h(\theta)|z]^2\right) ||f^t - f||_1^2 < C_h^2 \varrho^{2t}.$$

□□

Liu, Wong and Kong (1994) show that, when $(\theta^{(t)})$ satisfies a strong duality property, additional convergence properties, such as Rao-Blackwellization and monotone decrease of the covariances, are satisfied. More precisely, a Markov chain $(x^{(t)})$ is said to satisfy the *interleaving property* when there exists a second chain $(y^{(t)})$ such that

1. $x^{(t)}$ and $x^{(t+1)}$ are independent conditionally on $y^{(t)}$;

2. $y^{(t-1)}$ and $y^{(t)}$ are independent conditionally on $x^{(t)}$;

3. $(x^{(t)}, y^{(t-1)})$ and $(x^{(t)}, y^{(t)})$ are identically distributed under stationarity.

Both chains are then said to be *interleaved*. In the case of Data Augmentation, the existence of the chain $(z^{(t)})$ is thus sufficient to establish that the interleaving property is always satisfied. Note that the global chain $(z^{(t)}, \theta^{(t)})$ is not necessarily reversible, while both chains $(z^{(t)})$ and $(\theta^{(t)})$ are reversible. In fact, if K_1 is the transition kernel for $(z^{(t)})$ and $g(z, \theta) = \pi(\theta|z)f(z)$, the following *detailed balance condition*

$$f(z_0) K_1(z_0, z_1) = f(z_0) \int \pi(\theta_0|z_0) f(z_1|\theta_0) d\theta_0$$

$$= \int g(z_0, \theta_0) f(z_1|\theta_0) d\theta_0$$

$$= \int g(z_0, \theta_0) \frac{\pi(\theta_0|z_1) f(z_1)}{\pi(\theta_0)} d\theta_0$$

$$= f(z_1) K_1(z_1, z_0),$$

holds and, similarly, if K_2 denotes the kernel for $(\theta^{(t)})$,

$$\pi(\theta_0) K_2(\theta_0, \theta_1) = \pi(\theta_1) K_2(\theta_1, \theta_0) .$$

A first result of importance for convergence control is that the covariances $\text{cov}(h(\theta^{(1)}), h(\theta^{(t)}))$ are monotonically decreasing for every function h (see Liu *et al.*, 1994, or Robert and Casella, 1998, for a proof). This property is particularly handy in the estimation of the asymptotic variance for the Central Limit Theorem, because it allows to truncate the sum

$$\sum_{t=1}^{\infty} \text{cov}(h(\theta^{(1)}), h(\theta^{(t)}))$$

when $\text{cov}(h(\theta^{(1)}), h(\theta^{(t)}))$ is small enough.

Lemma 3.2.2 *If* $(\theta^{(t)})$ *is a Markov chain with the interleaving property, the covariances*

$$\text{cov}(h(\theta^{(1)}), h(\theta^{(t)}))$$

are positive and decreasing in t for every $h \in \mathcal{L}_2(\pi)$.

The result on the domination of the usual average by the Rao-Blackwellized version is based on a lemma which represents covariances via embedded expectations and which validates in addition *batch sampling* proposed in some versions of the MCMC algorithms (Diebolt and Robert, 1990; Geyer, 1992; Raftery and Lewis, 1992a). It also leads to the following validation of Rao-Blackwellization:

Theorem 3.2.3 *If* $(\theta^{(t)})$ *and* $(z^{(t)})$ *are interleaved Markov chains, the Rao-Blackwellized estimator improves on the usual average for every function* $h \in \mathcal{L}_2(\pi)$ $(i = 1, 2)$.

Note that some conditions need to be imposed for (3.1) to improve on the usual estimate, since Liu, Wong and Kong (1994) provide counterexamples where Rao-Blackwellization increases the variance. Interleaving is only a sufficient condition but Geyer (1995) has proposed a necessary and sufficient condition which is of practical interest only in finite setups. This condition is defined in terms of Hilbert operators in the space $\mathcal{L}_2(\pi)$: if T corresponds to the kernel operator, with $(Th)(y) = \mathbb{E}[h(x^{(t)})|x^{(t-1)} = y]$, if $A = (id - T)^{-1}$, A^* is the adjoint operator and if

$$B = A^*A - A^*T^*TA,$$

the necessary and sufficient condition is that the conditional expectation commutes with B. It is thus difficult to check in continuous setups.

Liu, Wong and Kong (1995) also extend Lemma 3.2.2 to the *randomized Gibbs sampler*, where each step only actualizes one component of y, chosen randomly according to a distribution $\sigma = (\sigma_1, \cdots, \sigma_p)$, even though interleaving does not hold. The transition from time t to time $t + 1$ is then

1. Select the component ν according to the distribution σ.

2. Generate $y_\nu^{(t+1)} \sim g_\nu(y_\nu | y_j^{(t)}, j \neq \nu)$ and take   [A_8]

$$y_j^{(t+1)} = y_j^{(t)} \qquad \text{for} \qquad j \neq \nu.$$

Note that this version of the Gibbs sampler is reversible.

Rao–Blackwellization is an elegant (and free from manipulation) method to approximate the densities of the different components of y, since

$$\frac{1}{T}\sum_{t=1}^{T} g_i(y_i|y_j^{(t)}, j \neq i)$$

is unbiased and converges to the marginal density $g_i(y_i)$. This property avoids the call for non-parametric methods, while being more accurate. This representation will be used in §3.3 to derive control variates for Gibbs samplers.

McKeague and Wefelmeyer (1995) propose a stronger although less practical conditioning on the previous value of the Markov chain and prove that the resulting Rao-Blackwellised estimator has asymptotically a smaller variance when the Markov chain is reversible. In cases this version of Rao-Blackwellization can be implemented, it dominates the one above by simple arguments following from Liu, Wong and Kong (1994). The applicability of this form of Rao-Blackwellization is nonetheless very restricted, with limited applicability outside precise finite setups such as Ising models. One could think that Metropolis–Hastings algorithms are naturally open to this kind of conditioning, but this is not usually the case. Moreover, the more radical conditioning of Casella and Robert (1996), which integrates out the uniform variables used in the Metropolis–Hastings acceptance step, is bound to lead to better performances, since it somehow mixes the sample more efficiently.

Example 3.2.1 Consider the transition kernel

$$x^{(t+1)} = \begin{cases} x^{(t)} & \text{with probability } 1 - x^{(t)}, \\ y \sim \mathcal{B}e(\alpha + 1, 1) & \text{otherwise,} \end{cases}$$

which is associated with the $\mathcal{B}e(\alpha, 1)$ as stationary distribution f (see Robert and Casella, 1998, Chapter 6, for motivations). For $h(x) = x^{1-\alpha}$, the conditional expectation $\mathbb{E}[h(x^{(t+1)})|x^{(t)}]$ is given by

$$\begin{aligned} \mathbb{E}[h(x^{(t+1)})|x^{(t)}] &= (1 - x^{(t)})h(x^{(t)}) + x^{(t)}\mathbb{E}[y^{1-\alpha}] \\ &= (1 - x^{(t)})h(x^{(t)}) + x^{(t)}(\alpha + 1)\int_0^1 y^{1-\alpha+\alpha}dy \\ &= (1 - x^{(t)})h(x^{(t)}) + x^{(t)}(\alpha + 1)/2, \end{aligned}$$

which leads to the following Rao-Blackwellised estimate of $\mathbb{E}^f[x^{1-\alpha}]$:

$$\frac{1}{T}\sum_{t=1}^{T}\left\{(1 - x^{(t)})h(x^{(t)}) + x^{(t)}(\alpha + 1)/2\right\}.$$

Note that the procedure can be iterated in this case, namely that the computation of $\mathbb{E}[h(x^{(t+1)})|x^{(t-1)}] = \mathbb{E}[\mathbb{E}[h(x^{(t+1)})|x^{(t)}]|x^{(t-1)}]$ and of the following conditional expectations is feasible. However, the improvement brought by the Rao-Blackwellization is negligible in the sense that the corresponding estimate is indistinguishable from the usual empirical average (see Robert, 1995). ‖

The setup of Example 3.2.1 is particularly artificial, given that we can compute the expectation $\mathbb{E}[x^{1-\alpha}]$ and even simulate directly the stationary distribution. In more realistic settings, the expectation

$$
\begin{aligned}
\mathbb{E}[h(x^{(t+1)})|x^{(t)}] &= \int h(y)\rho(x^{(t)}, y)q(y|x^{(t)})dy \\
&+ h(x^{(t)}) \int (1 - \rho(x^{(t)}, y))q(y|x^{(t)})dy
\end{aligned}
$$

involves integrals of the form

$$
\int \left\{ \frac{f(y)}{f(x^{(t)})} \frac{q(x^{(t)}|y)}{q(y|x^{(t)})} \wedge 1 \right\} q(y|x^{(t)})dy ,
$$

which are usually impossible to compute.

One may wonder about the relevance of Rao-Blackwellization for convergence control since it seems that, at least in Gibbs sampling setups, the improvement brought by Rao-Blackwellization is most often negligible (see Robert, 1995). This is not the case, though, for the "non-parametric" Rao-Blackwellization of Casella and Robert (1996) which eliminates the generic uniform random variables used in the algorithms by explicit integration. An additional incentive in using Rao-Blackwellization for convergence control is to select specific functions h, akin to conjugate priors in Bayesian analysis, whose role is to provide explicit expectations in (3.1). This idea is developed in full detail in §3.3.

Example 3.2.2 In a logistic regression model, we observe $x_1, \ldots, x_m$ with $x_i \sim \mathcal{B}(n_i, p_i)$ and

$$
p_i = p_i(\alpha) = \frac{\exp(\alpha t_i)}{1 + \exp(\alpha t_i)}, \qquad 1 \leq i \leq m,
$$

where the t_i's are fixed (or the model is conditional on the t_i's). Consider conjugate priors of the form ($\lambda > 0$)

$$
\pi(\alpha|y_0, \lambda) \propto \exp(\alpha y_0) \prod_{i=1}^{m} (1 + e^{\alpha t_i})^{-\lambda n_i} \tag{3.2}
$$

Indeed, since a sufficient statistic is $y = \sum t_i x_i$, the posterior distributions are $\pi(\alpha|y_0 + y, \lambda + 1)$. Since simulation from (3.2) is impossible, a particular

MCMC algorithm is based on the random walk $\beta \sim \mathcal{N}(\alpha^{(n)}, \sigma^2)$, with $\alpha^{(n+1)} = \beta$ with probability

$$\frac{\pi(\beta|y_0 + y, \lambda + 1)}{\pi(\alpha^{(n)}|y_0 + y, \lambda + 1)} \wedge 1,$$

and $\alpha^{(n+1)} = \alpha^{(n)}$ otherwise. An interesting feature of this example is that the average

$$\overline{\psi'}_N = \frac{1}{N}\sum_{n=1}^{N}\psi'(\alpha^{(n)}) = \frac{1}{N}\sum_{n=1}^{N}\left(\sum_{i=1}^{m}t_i n_i p_i(\alpha^{(n)})\right),$$

has known expectation, namely the maximum likelihood estimator y. Therefore, it is possible to monitor the convergence of the algorithm via the convergence of (3.8) to y, i.e. by simply stopping the algorithm when the difference between (3.8) and y is small enough.

This procedure is illustrated for the data from Jensen $et\ al.$ (1991), given in Table 3.1, where $y = 43$, for $\lambda = 1$ and $y_0 = 0.0$. Figure 3.1 and Table 3.2 show the result of the experiment for three values of σ, 0.1, 1 and 10, leading to acceptance rates of .86, .28 and .03 respectively. As forecasted by the theoretical results of Gelman, Gilks and Roberts (1996), the Metropolis–Hastings algorithm with the acceptance rate of .28 does much better than for the two other variances, both in terms of stability of the estimates and in rapid occurrence of the stopping rule. Table 3.2 illustrates the lack of complete warranty of this empirical criterion, namely that the chain may stop too early to have achieved a sufficiently small variance. Note however that, for the precision .001, the estimates associated with $\sigma = 1$ are much closer to the limiting values than for the two other choices. ||

FIGURE 3.1. Convergence paths for the three estimates of $\mathbb{E}[\alpha|y]$ $(bottom)$ and control curves for $|\overline{\psi'}_N - 43|$ (top).

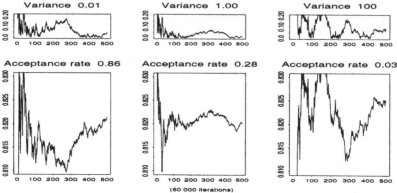

TABLE 3.1. Binomial data from Jensen *et al.* (1991).

i	t_i	n_i	x_i
1	-1	15	4
2	1	30	23
3	2	15	12

TABLE 3.2. MCMC estimates obtained by the control of the convergence of (3.8) through the stopping rules $|\overline{\psi'}_N - 43| < .01$ *(first line)* and $|\overline{\psi'}_N - 43| < .001$ *(second line)*. These estimates are to be compared with the limiting values 0.8195 and 0.3295 for the posterior expectation of α and the posterior probability that α is larger than 0.916 respectively, obtained after $500,000$ iterations for $\sigma = 1$.

σ	0.1	1.0	10.0
N	309	166	2548
	(1326)	(873)	(2552)
$\mathbb{E}[\alpha|y]$	.8101	.8276	.8203
	(.8224)	(.8187)	(.8209)
$P(\alpha > 0.916)$	.2686	.4398	.2465
	(.3401)	(.3356)	(.2476)

3.3   Riemann sum control variates

3.3.1   Merging numerical and Monte Carlo methods

In §3.2, the Rao-Blackwell estimator illustrates the improvement brought by the application of a frequentist principle in a Monte Carlo setup. Another type of improvement can be derived through the use of numerical techniques, which provides potentially new perspectives to improve upon standard Monte Carlo estimators. In this spirit, Philippe (1997a) proposes to consider Riemann sums, that is trapezoidal approximations of integrals, and to replace the standard uniform partition by a partition based on independent random variates generated from the distribution of interest. More precisely, the Riemann sum approximation to the integral,

$$\int_a^b h(x)f(x)dx \simeq \sum_{i=1}^n (a_{i+1} - a_i)h(\xi_i)f(\xi_i),$$

with $a_1 = a \le a_1 \le \ldots \le a_{n+1} = b$ and $\xi_i \in (a_i, ai + 1)$, can be replaced by the following sum:

Definition 3.3.1 *Consider an independent sample $(x^{(1)}, \cdots, x^{(T)})$ distributed from f and define $x^{(1:T)} \le \ldots \le x^{(T:T)}$ as the corresponding ordered* *sample. The* Riemann estimator *of the integral* $\mathbb{E}^f[h]$ *is*

$$\delta_T^R = \sum_{t=1}^{T-1} \left(x^{(t+1:T)} - x^{(t:T)}\right) h\left(x^{(t:T)}\right) f\left(x^{(t:T)}\right). \qquad (3.3)$$

When the density f is known up to a constant, c, the value of c can be approximated by using $h = 1$ in (3.3). As shown by Philippe (1997b), the sample $(x^{(1)}, \cdots, x^{(T)})$ can also be generated from an *instrumental density* g which has the same support as f, in the spirit of importance sampling. Note that, in opposition to classical importance sampling methods, the density g needs not be known, which may be of considerable interest in some setups. The convergence properties given in Proposition 3.3.1 show that the Riemann estimator (3.3) dominates the empirical average in terms of the speed of convergence of their variance (see Philippe, 1997a,b, for a proof).

Proposition 3.3.1 *The Riemann estimator (3.3) associated with the instrumental density g satisfies the following properties:*

1. If $hf/g \in \mathcal{L}^1(g)$ then

$$\lim_{T\to\infty} \mathbb{E}\left[\delta_T^R\right] = \mathbb{E}^f[h].$$

Moreover, if the ratio hf/g is bounded then the convergence rate of the bias is $O(T^{-1})$.

2. If $hf/g \in \mathcal{L}^2(g)$ then

$$\lim_{T\to\infty} \mathbb{E}\left[\left(\delta_T^R - \mathbb{E}\left[\delta_T^R\right]\right)^2\right] = 0.$$

Moreover if the ratio hf/g and its derivative $(hf/g)'$ are bounded on the support of f then the convergence rate of the variance is $O(T^{-2})$.

The Riemann estimator is thus an efficient alternative to the classical Monte Carlo estimator for independent samples since it converges faster (by an order of magnitude). Philippe and Robert (1998) show in addition that the convergence properties are preserved when we consider Riemann sums based on the output of convergent Markov Chain Monte Carlo algorithms.

Example 3.3.1 (Example 1.3.2 cont.) Consider a sample from the density

$$f(\theta|\theta_0) \propto \frac{e^{-\theta^2/2}}{[1 + (\theta - \theta_0)^2]^\nu}, \qquad (3.4)$$

simulated by a Data Augmentation method, based on the following conditional densities

$$\eta|\theta \sim \mathcal{G}a\left(\nu, \frac{1 + (\theta - \theta_0)^2}{2}\right), \qquad \theta|\eta \sim \mathcal{N}\left(\frac{\theta_0\eta}{1+\eta}, \frac{1}{1+\eta}\right),$$

as detailed in Example 1.3.2. Since the density $f(\cdot|\theta_0)$ is known up a multiplicative constant, the Riemann sums can be used. For instance, if $h(\theta) = \theta$, the Riemann sum approximation to $\mathbb{E}[\theta]$ is

$$\delta_T^R = \frac{\sum_{t=1}^{T-1}(\theta^{(t+1:T)} - \theta^{(t:T)})\theta^{(t:T)}e^{-\theta^{(t:T)^2}/2}[1 + (\theta^{(t:T)} - \theta_0)^2]^{-\nu}}{\sum_{t=1}^{T-1}(\theta^{(t+1:T)} - \theta^{(t:T)})e^{-\theta^{(t:T)^2}/2}[1 + (\theta^{(t:T)} - \theta_0)^2]^{-\nu}}.$$

where $\theta^{(1:T)} \leq \cdots \leq \theta^{(T:T)}$ are the ordered values of the Markov chain. Since the expectation of the distribution of θ conditional on η is available, we can compare the performance of the Riemann sums with the Rao-Blackwell estimator of $\mathbb{E}[\theta]$, which is equal to

$$\delta_T^{Rb} = \sum_{t=1}^{T} \frac{\theta_0 \eta^{(t)}}{1 + \eta^{(t)}}$$

in this case, and to the empirical average as well. Figure 3.2 illustrates the considerable improvement brought by using the Riemann estimator in terms of stability and convergence rate to the true value. Moreover, we ran a Monte Carlo experiment based on 2000 replications, in order to evaluate a two-sided α–credible region, that is, a region C_T^α such that, for a given sample size T,

$$P(\delta_T \in C_T^\alpha) = \alpha .$$

The contrasted amplitudes of the different regions clearly show that the improvement brought by the Riemann estimator upon both the empirical average and the Rao-Blackwell estimator is far from negligible.   ‖

FIGURE 3.2. Convergence paths for the three estimates of $\mathbb{E}^f[\theta]$: empirical average *(plain)*, Riemann estimator *(dots)* and Rao-Blackwell estimator *(dashes)* *(left)* comparison of the 95%–credible region *(right)* (for $\nu = 2$ and $\theta_0 = 1$ in (3.4)).

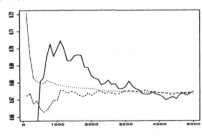 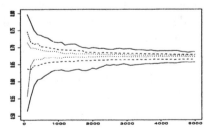

3.3.2   Rao-Blackwellized Riemann sums

When the random variable x to be simulated is a vector, the Riemann sum method does not apply so readily, in the sense that a grid must replace the interval partition provided by the order statistics and, most dramatically, that the performances of the associated estimator considerably drop[2] with the dimension p, to the point that it is not an acceptable competitor to the standard empirical average when $p \geq 3$. In this set-up, Philippe and Robert (1998) propose an alternative which associates the Rao-Blackwellization technique with Riemann sums. The method is based on the Bayes decomposition of the density of x as the conditional density of a component x_I multiplied by the marginal density of the remaining components, $x_{\backslash I} = (x_1, \cdots, x_{I-1}, x_{I+1}, \cdots x_p)$ $(I = 1, \ldots, p)$,

$$f(x) = \pi(x_I | x_{\backslash I}) \pi(x_{\backslash I}) .$$

Therefore, the expectation $\mathbb{E}^f[h(x)]$ can be decomposed as

$$\int \int h(x) \pi(x_I | x_{\backslash I}) \pi(x_{\backslash I}) dx_{\backslash I} dx_I$$

and the approximative integration operates in two steps. First, for a fixed value x_I, the integral in $x_{\backslash I}$ is replaced with its Rao-Blackwellized approximation, that is

$$\frac{1}{T} \sum_{k=1}^{T} h\left(x_{\backslash I}^{(k)}, x_I\right) \pi\left(x_I | x_{\backslash I}^{(k)}\right) ,$$

which converges to

$$\phi(x_I) = \int h\left(x_{\backslash I}, x_I\right) \pi\left(x_I | x_{\backslash I}\right) \pi\left(x_{\backslash I}\right) dx_{\backslash I} .$$

This argument allows to eliminate the multidimensional integration and to fall back on an unidimensional problem, namely the integration of $\phi(x_I)$ against x_I. The Riemann sum approximation thus applies and proposes

$$\delta_T^I = T^{-1} \sum_{t=1}^{T-1} (x_I^{(t+1:T)} - x_I^{(t:T)}) \left\{ \sum_{k=1}^{T} h\left(x_I^{(t:T)}, x_{\backslash I}^{(k)}\right) \pi^I\left(x_I^{(t:T)} | x_{\backslash I}^{(k)}\right) \right\}$$

$$(3.5)$$

as an estimator of the integral of ϕ.

When all the conditional densities $\pi\left(x_I | x_{\backslash I}\right)$ are available, note that this approach produces p different estimators of the integral $\mathbb{E}^f[h(x)]$, which are all based on the same output of the Markov chain. A first use of this

---

[2] This is a classical occurrence of the "curse of dimensionality" in numerical analysis.

variety of estimates is to try to identify the fastest, in order to reduce the convergence time. A second application, more oriented towards convergence monitoring, is to call for the graphical tool introduced in §2.2.1, by comparing the p Rao-Blackwellised Riemann sums till they all take similar values. Obviously, the superior performances of the genuine Riemann sum are not preserved, because the speed of convergence is dictated by the speed of convergence of the Rao-Blackwellised estimator of the conditional density. The next section focus on this second application to propose a specific control variate.

3.3.3   Control variates

While the convergence of the different Markov Chain Monte Carlo estimators (3.5) to the same value is a necessary condition, it is not sufficient to assess whether the chain has completely explored the support of f, especially when the monitoring is based on a single path of the Markov chain. We now show how a single chain can still produce an evaluation of the "missing mass", that is of the part of the support of f which has not yet been explored by the chain.

First, when the integration problem is unimodal, the estimator (3.3) can be used with the constant function $h(x) = 1$. In this special case, (3.3) works as a control variate in the sense that it must converge to 1 for the chain to converge. Moreover, this provides us with an "on-line" evaluation of the probability of the region yet unexplored by the chain.

In the general case, the Rao-Blackwellised Riemann sum extension (3.5) can replace (3.3) for the constant function $h(x) = 1$, that is ($1 \le I \le p$)

$$\Delta_T^I = T^{-1} \sum_{t=1}^{T-1} \left(y_I^{(t+1:T)} - y_I^{(t:T)} \right) \left(\sum_{k=1}^{T} \pi \left(y_I^{(t:T)} | y_{\backslash I}^{(k)} \right) \right) . \tag{3.6}$$

In this estimate, the average

$$T^{-1} \sum_{k=1}^{T} \pi \left(y_I^{(t:T)} | y_{\backslash I}^{(k)} \right) \tag{3.7}$$

corresponds to the Rao-Blackwell estimation of the marginal density of the I-th component. Therefore, the quantity (3.6) converges to

$$\int \pi(y_I) \, dy_I = 1,$$

and also works as a control variate as in the univariate case.

While the speeds of convergence of the Rao-Blackwell estimators (3.7) may differ (in I), a requirement for a good exploration of the support of f is to impose that all Δ^I are close to 1. Moreover, the different speeds of convergence can help in identifying the components which are more slowly mixing and in suggesting alternative instrumental distributions. The drawback with this method is that it fails to explore multidimensional aspects of the joint density and also that it does not apply with large numbers of iterations, because of the time requirement related to sorting the Markov Chain Monte Carlo output.

FIGURE 3.3. Convergence paths for the three estimates of $\mathbb{E}^f[\theta]$ *(top)* and control variate curves *(bottom)* for the Cauchy Benchmark.

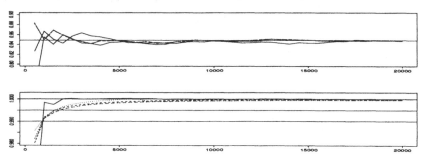

Example 3.3.2 As shown in Example 2.2.1, the full conditional densities for the completed model are given by

CAUCHY
BENCHMARK

$$\eta_i|\theta, x_i \sim \mathcal{E}xp\left(\frac{1 + (\theta - x_i)^2}{2}\right), \qquad (i = 1, 2, 3)$$

$$\theta|x_1, x_2, x_3, \eta_1, \eta_2, \eta_3 \sim \mathcal{N}\left(\frac{\eta_1 x_1 + \eta_2 x_2 + \eta_3 x_3}{\eta_1 + \eta_2 + \eta_3 + \sigma^{-2}}, \frac{1}{\eta_1 + \eta_2 + \eta_3 + \sigma^{-2}}\right).$$

Therefore, it is possible to propose four different estimates of the parameter of interest, θ, using the same simulated Markov chain. Figure 3.3 illustrates the convergence of the corresponding Rao-Blackwellised Riemann sums. It shows that the Δ_T^I's in (3.5) enjoy different convergence speeds, in the sense that the control variate estimates (3.6) read the 1% or .5% error band for different numbers of iterations. Nonetheless, the convergence to 1 occurs quite rapidly for the four estimates. In addition, the four control variates converge to 1 from below, thus exhibiting the initial mass loss of the Markov chain and the good performances of the Rao-Blackwell estimators of the marginal densities.   ‖

Example 3.3.3 As in the above example, the full posterior distributions,

PUMP
BENCHMARK

$$\lambda_i|\beta, t_i, p_i \sim \mathcal{G}a(p_i + \alpha, t_i + \beta), \qquad (1 \le i \le 10),$$

$$\beta | \lambda_1, \cdots, \lambda_{10} \quad \sim \quad \mathcal{G}a\left(\gamma + 10\alpha, \delta + \sum_{i=1}^{10} \lambda_i\right),$$

are completely known. Therefore, we can now construct eleven different Rao-Blackwellised Riemann sum estimates of β. Figure 3.4 illustrates the convergences of these different estimates, as well as the convergence of the control variates. In contrast with Example 3.3.2, note that some control variates converge to 1 from above, which indicates that the convergence of the Rao-Blackwell density estimators occurs more slowly than the recovery of the missing probability mass by the Markov chain.   ‖

FIGURE 3.4. Convergence paths for the different Rao Blackwellised Riemann sums of $\mathbb{E}^f[\beta]$ *(top)* $\mathbb{E}[\lambda]$ *(middle)* and control curves *(bottom)* for the Pump Benchmark.

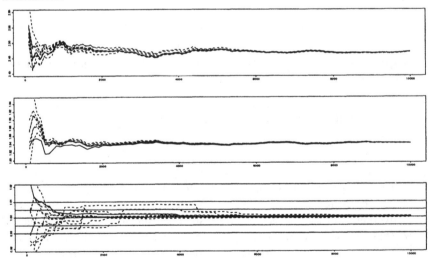

3.4 A mixture example

Consider a two-component normal mixture[3] distribution

$$p\mathcal{N}(\theta_1, \sigma_1^2) + (1 - p)\mathcal{N}(\theta_0, \sigma_0^2), \qquad (3.8)$$

with the conjugate prior distributions

$$p \sim \mathcal{B}e(1/2, 1/2), \quad \theta_i \sim \mathcal{N}(\xi_i, \sigma_i^2/n_i), \quad \sigma_i^2 \sim \mathcal{IG}(\nu_i/2, \omega_i^2/2) \qquad (i = 1, 2).$$

Given a sample $x_1, \ldots, x_n$ from (3.2), the posterior distribution appears as a sum of 2^n closed form terms from exponential families and this combinatoric explosion requires an MCMC approximation when n is larger than 30 (see Diebolt and Robert, 1990, 1994). The fruitful approach to the mixture problem is to perceive the model as a *missing data structure*, by introducing dummy variables $z_1, \ldots, z_n$, which indicate the components from which the x_i's originated. The 'completed model' stands as follows:

$$P(z_i = 1) = 1 - P(z_i = 2) = p, \qquad x_i | z_i \sim \mathcal{N}(\theta_{z_i}, \sigma_{z_i}^2),$$

and the corresponding Gibbs implementation is to simulate iteratively the missing data and the parameters as follows $(i = 1, \ldots, n)$:

1. Simulate

$$z_i \sim \mathcal{B}\left(\left\{1 + \frac{(1 - p)\sigma_1 \exp\left\{-(x_i - \theta_0)^2/2\sigma_0^2\right\}}{p\sigma_0 \exp\left\{-(x_i - \theta_1)^2/2\sigma_1^2\right\}}\right\}^{-1}\right)$$

Each simulation of the missing data provides two subsamples of sizes ω and $n - \omega$ corresponding to each component and related averages $\bar{m}_1$ and $\bar{m}_0$, sums of squared errors s_1^2 and s_0^2, i.e.

$$\omega = \sum_{i=1}^{n} \mathbb{I}_{z_i=1},$$

$$\omega\bar{m}_1 = \sum_{i=1}^{n} \mathbb{I}_{z_i=1} x_i,$$

$$s_1^2 = \sum_{i=1}^{n} \mathbb{I}_{z_i=1}(x_i - \bar{m}_1)^2.$$

The second step of the Gibbs iteration is then

---

[3] Chapter 8 extends on the estimation of mixtures of distributions by presenting additional convergence control methods in the setup of exponential distributions. See also §5.6.

2. Simulate $[A_9]$

(i) $p \sim \mathcal{B}e(\omega + 1, n - \omega + 1)$;

(ii) $\sigma_0^2 \sim \mathcal{IG}\left(\frac{\nu_0 + n - \omega}{2}, \frac{1}{2}\left\{\omega_0^2 + s_0^2 + \frac{n_0(n-\omega)}{n_1 + n - \omega}[s_0^2 + (n - \omega)(\bar{m}_0 - \xi_0)^2]\right\}\right)$

(iii) $\sigma_1^2 \sim \mathcal{IG}\left(\frac{\nu_1 + \omega}{2}, \frac{1}{2}\left\{\omega_1^2 + s_1^2 + \frac{n_1 \omega}{n_1 + \omega}[s_1^2 + \omega(\bar{m}_1 - \xi_1)^2]\right\}\right)$

(iv) $\theta_0 \sim \mathcal{N}\left(\frac{n_0 \xi_0 + (n-\omega)\bar{m}_0}{n_0 + n - \omega}, \frac{\sigma_0^2}{n_0 + n - \omega}\right)$.

(v) $\theta_1 \sim \mathcal{N}\left(\frac{n_1 \xi_1 + \omega \bar{m}_1}{n_1 + \omega}, \frac{\sigma_1^2}{n_1 + \omega}\right)$;

As a particular case of Gibbs sampling with strong aperiodicity and irreducibility at both stages, $[A_9]$ has the usual convergence properties. The dual structure of $[A_9]$ is quite obvious and shows that the Duality Principle also applies in this setting, relating the convergence of the Markov chain of interest $(p^{(t)}, \theta_1^{(t)}, \theta_0^{(t)}, \sigma_1^{(t)}, \sigma_0^{(t)})$ to those of a finite state space Markov chain $(z^{(t)})$. In this case, the state space is of cardinality 2^n, which may be an hindrance in the practical study of the convergence of the chain $(z^{(t)})$ itself. From a theoretical point of view, both chains are geometrically ergodic and the Central Limit Theorem applies.

Note that this property does not hold for alternative implementations of the Gibbs sampler. For instance, consider Mengersen and Robert's (1996) reparameterization

$$p\mathcal{N}(\mu, \tau^2) + (1 - p)\mathcal{N}(\mu + \tau\theta, \tau^2\sigma^2),$$

with

$$\pi(p, \mu, \tau) = \tau^{-1}, \quad \sigma \sim \mathcal{U}_{[0,1]} \quad \text{and} \quad \theta \sim \mathcal{N}(0, \zeta^2).$$

This equivalent representation of (3.2) express the parameters of the second component as a local perturbation of an overall location–scale parameter (μ, τ) and is mainly of interest in noninformative settings since it allows for improper priors on (μ, τ), in the sense that the posterior distribution is always finite (see Robert and Titterington, 1998). However, although it provides an higher efficiency in the Gibbs sampler in practice, this perspective requires full conditional distributions and the parameter $(p, \theta_1, \theta_0, \sigma_1, \sigma_0)$ is not generated conditionally on z. In fact, Step 2 in $[A_9]$ is replaced by

2. Simulate $[A_{10}]$

(i) $p \sim \mathcal{B}e(\omega + 1, n - \omega + 1)$;

(ii) $\sigma_0^{-2} \sim \mathcal{IG}\left(\frac{n - \omega - 2}{2}, \frac{(n - \omega)(\bar{m}_0 - \theta_0)^2 + s_0^2}{2}\right)\mathbf{I}_{[0,\sigma_1]}(\sigma_0)$;

(iii) $\sigma_1^{-2} \sim \mathcal{IG}\left(\frac{\omega + 2}{2}, \frac{\omega(\bar{m}_1 - \theta_1)^2 + s_1^2 + (\theta_0 - \theta_1)^2 \zeta^{-2}}{2}\right)\mathbf{I}_{[\sigma_0, \infty)}(\sigma_1)$;

(iv) $\theta_0 \sim \mathcal{N}\left(\frac{(n - \omega)\bar{m}_0 + \zeta^{-2}\theta_1 \sigma_0^2/\sigma_1^2}{n - \omega + \zeta^{-2}\sigma_0^2/\sigma_1^2}, \frac{\sigma_0^2}{n - \omega + \zeta^{-2}\sigma_0^2/\sigma_1^2}\right)$;

(v) $\theta_1 \sim \mathcal{N}\left(\frac{\omega\overline{m}_1 + \zeta^{-2}\theta_0}{\omega + \zeta^{-2}}, \frac{\sigma_1^2}{\omega + \zeta^{-2}}\right)$;

when expressed in the parameterization of (3.2). Therefore, given the dependence on the previous value of the parameter, it is not possible to use the finite state space chain $(z^{(t)})$ to create renewal sets. Moreover, the subchain $(z^{(t)})$ cannot be considered independently from the parameter subchain since *it is not a Markov chain*. The possibilities for convergence assessment are therefore reduced.

Robert, Celeux and Diebolt (1993) extend the MCMC treatment of mixture models to *hidden Markov chains*, which allow for a possible Markov dependence between the observations, $x_1, \ldots, x_n$, which can be described at the missing data level. The simulation of this missing data then gets too time-consuming to be operated directly and this imposes the following Gibbs decomposition:

1. Simulate $z_1^{(t)}|z_2^{(t-1)}, \ldots, z_n^{(t)}, \theta^{(t-1)}$;

$$\ldots \tag{A_{11}}$$

n. Simulate $z_n|z_1^{(t)}, \ldots, z_n^{(t)}, \theta^{(t-1)}$,

while $\theta^{(t)} \sim \pi(\theta|z^{(t)})$ as in $[A_{10}]$. This type of decomposition implies that $(\theta^{(t)})$ is not a Markov chain since $z^{(t)}$ is generated from a distribution of the form $f(z|\theta^{(t)}, z^{(t-1)})$. It illustrates once again that the Duality Principle applies in a wider context than just Data Augmentation. For instance, the finiteness of the state space of $z = (z_1, \ldots, z_n)$ and the irreducibility of the Markov chain $(z^{(t)})$ ensure that the Central Limit Theorem holds. (See Robert and Titterington, 1998, for a parameterization of hidden Markov models which extends Mengersen and Robert, 1996, and for which the Duality Principle does not hold.)

Convergence assessment in this setup is, to say the least, an open problem because of the complex structure of the posterior distribution. Indeed, when developing the likelihood in a sum of 2^n normal-type terms, it is easy to see that the posterior distribution also looks like a weighted sum of standard conjugate distributions and is thus most likely multimodal. When considering a k component mixture,

$$\sum_{i=1}^{k} p_i \mathcal{N}(\theta_i, \sigma_i^2),$$

the complexity of the posterior increases exponentially, with significant modes around k' component submodels ($k' < k$). The extension of $[A_9]$ is then unlikely to explore the whole range of the posterior modes and usually stays in the neighborhood of one of the major modes, with rare jumps between modes. Obviously, most of the modes are minor and some

FIGURE 3.5. Evolution of the MCMC estimate of the mixture density as the number of iterations T increases for the stamp dataset of Izenman and Sommer (1988) with $k = 3$ and $\zeta^2 = 0.01$.

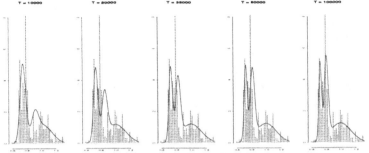

others are redundant, being permutation transforms of the main modes, but it is impossible to exclude some modes from the start.

To illustrate this difficulty in exploring the posterior surface, we examine a dataset which was first analyzed by Izenman and Sommer (1988). It consists of the measurements of the thickness of 485 Mexican stamps from the 1872 Hidalgo issue. For very convoluted reasons related to the different types of papers used in the printing of stamps, the thickness is heterogeneous with historical validations of mixtures from 3 to 7 components. A straightforward application of the likelihood ratio test led Izenman and Sommer (1988) to propose a 3 component mixture model, while the non-parametric kernel estimation supported a seven mode density. A reappraisal of the data by Basford, McLachlan and York (1998) is also in favor of a normal mixture of seven components *with equal variances*. Another analysis in Robert and Mengersen (1998) does not give any reason to prefer $k = 4, 5, 6, 7$ over $k = 3$, thus concurring to the original solution of Izenman and Sommer (1988). Figure 3.5 shows the succession of the estimates of the mixture density as the number of iterations in the Gibbs sampler $[A_{10}]$ increases to 100,000, exhibiting the progressive appearance of a bimodal structure in the first part of the density.

The most trivial assessments do not work well in this setup. For instance, simple plots of the convergence curves for the parameters never seem to settle, as seen on Figure 3.6, even though the corresponding estimates of the densities look stable (see Figure 3.7). This phenomenon exhibited in Robert and Mengersen (1998) can be blamed on the weak identifiability of mixture models where different sets of parameters $(p_i, \theta_i, \sigma_i)$ may produce almost identical distributions. Similarly, although Rao–Blackwellisation is feasible for the algorithm $[A_9]$ and even for some components of $[A_{10}]$, the comparison of usual estimates and Rao-Blackwell corrections usually brings very little information as whether or not convergence has been achieved since both classes of estimates are indistinguishable from the start (see Robert and Mengersen, 1998, for examples).

FIGURE 3.6. Evolution of the MCMC estimate of some parameters of the 3 component mixture corresponding to Figure 3.5. (*Source:* Robert and Mengersen, 1998.)

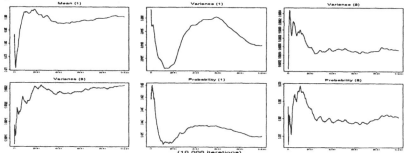

FIGURE 3.7. Estimates of the density associated to successive estimates of the first variance, as in Figure 3.6.

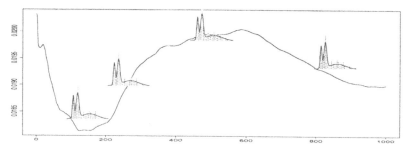

A more advanced control of convergence must then be achieved through the finite state chain $(z^{(t)})$ of the allocations, although the state space is then of dimension k^n. Figure 3.8 provides a description of the evolution of the allocations for the 485 points of the sample, each component being represented by a different grey level. (Robert, 1997, calls this representation an *allocation map*.) The image on the left corresponds to 5000 iterations of the algorithm $[A_{10}]$, with a random start and a large precision $\zeta^{-1} = 1$. The sudden modification on the upper part of the image shows that stationarity has not been achieved. On the contrary, the stable bands of the image on the right are indicative of a strong stability. The corresponding mixture estimates on top of both images confirm this assessment. (Note that the scale of the allocation maps (observation index and iteration number) differ from the scale of the histograms [observation value and probability].) However, a word of caution must be added at this point: the stability of the allocation map is neither necessary nor sufficient for convergence control. Some setups, like ill-separated components (Chapter 8) or mixtures with an unknown number of components (Richardson and Green, 1997, Robert, 1997, Gruet *et al.*, 1998) may produce allocations maps which "never" stabilise. In addition, stable allocations as in Figure 3.7 *(right)* indicate that the algorithm has presumably found a mode of the posterior distribution, but also that it is unable to leave the neighbourhood of this mode and therefore that its mixing properties may be unsatisfactory.

FIGURE 3.8. Image representations of the successive allocations of the 485 stamps of the 1872 Hidalgo issue sample to the components of a 3 component mixture representation. The shades correspond to the three components *(white is for 1, grey for 3 and dark for 2)*. The estimations above are those obtained after averaging over 5000 *(left)* and 50,000 *(right)* iterations.

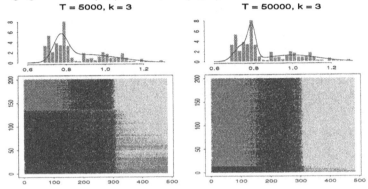

4
Valid Discretization via Renewal Theory

Chantal Guihenneuc–Jouyaux
Christian P. Robert

4.1 Introduction

As discussed in Chapter 2, an important drawback of Raftery and Lewis'
(1992a, 1996) convergence control method is that the discretized version of
the Markov chain is not a Markov chain itself, unless a stringent *lumpability*
condition holds (see Kemeny and Snell, 1960). This somehow invalidates the
binary control method, although it provides useful preliminary information
on the required number of iterations. However, the discrete aspect of the
criterion remains attractive for its intuitive flavour and, while the Duality
Principle of Chapter 1 cannot be invoked in every setting, this chapter
shows how *renewal theory* can be used to construct a theoretically valid
discretization method for general Markov chains. We then consider some
convergence control methods based on these discretized chains, even though
the chains can be used in many alternative ways (see also Chapter 5).

Section 4.2 contains the essentials of renewal theory, Section 4.3 describes
the discretization method and Section 4.4 studies a particular evaluation,
based on Guihenneuc–Jouyaux and Robert (1998), while Section 4.6 men-
tions an alternative use of renewal theory. Note that Mykland, Tierney
and Yu (1995) and Gilks, Roberts and Sahu (1998) have also studied the
implications of renewal theory on the control of MCMC algorithms, while
Athreya, Doss and Sethuraman (1996) show how it can justify the theoretical derivation of convergence results for these algorithms.

4.2 Renewal theory and small sets

4.2.1 Definitions

As noted by Mykland, Tierney and Yu (1995) and Robert (1995), the renewal properties of the Markov chain under study can be used to assess
convergence of the chain to the stationary distribution and to improve the

estimation of the parameters of this distribution. From our point of view (of monitoring the convergence of MCMC algorithms), the main appeal of renewal theory is that, when it applies, the study of the generic sums

$$S_T = \sum_{t=1}^{T} h(x^{(t)})$$

can be simplified in a monitoring of iid random variables and a classical form of the Central Limit Theorem then applies. This transformation to a simpler setting is actually done by decomposing S_T into a sum of iid random variables. (See Orey, 1971, Lindvall, 1992, Meyn and Tweedie, 1993, and Athreya *et al.*, 1996, for deeper theoretical treatments.)

A condition for renewal theory to apply is that there exist a set A, a real $0 < \epsilon < 1$ and a probability measure ν such that $\nu(A) > 0$ and

$$\forall x^{(t)} \in A, \ \forall B, \qquad P(x^{(t+1)} \in B | x^{(t)}) \geq \epsilon\nu(B). \qquad (4.1)$$

(For simplicity's sake, we assume that the chain is strongly aperiodic. Otherwise, we would need to define the notion for an m-batch subchain of $(x^{(t)})$, $(x^{(tm)})$.) The set A is called *renewal set* (Asmussen, 1979) or *small set* (Meyn and Tweedie, 1993). It can be shown that small sets exist for the chains involved in MCMC algorithms since it follows from Asmussen (1979) that every irreducible Markov chain allows for renewal. Meyn and Tweedie (1993) also show that the whole space can be covered with small sets. The practical determinations of small sets and of the corresponding (ϵ, ν) are more delicate but Mykland *et al.* (1995) and Robert (1995) have shown that this can be done in realistic setups, sometimes through a modification of the transition kernel. In discrete cases, A can be selected as the collection of the most frequent states of the chain, based on either the transition matrix $\mathbb{P}$ or on a preliminary run. The bounding measure ν is then derived as

$$\nu(E) = \inf_{x^{(t)} \in A} P(x^{(t+1)} \in E | x_{(t)}).$$

(See also §4.5 for examples related to the Gibbs sampler.)

4.2.2 Renewal for Metropolis–Hastings algorithms

The approach of Mykland *et al.* (1995), sometimes called *regeneration*, extends (4.1) by replacing small sets with functions s such that

$$P(x^{(t+1)} \in B | x^{(t)}) \geq s(x^{(t)}) \, \nu(B), \qquad x^{(t)} \in \Theta, \ B \in \mathcal{B}(\mathcal{X}) . \qquad (4.2)$$

Small sets are then a special case of (4.2), with $s(x) = \epsilon\mathbb{I}_C(x)$. The probability of regeneration at each step is then

$$r(x^{(t)}, \xi) = \frac{s(x^{(t)}) \, \nu(\xi)}{K(x^{(t)}, \xi)} ,$$

where $K(x, \cdot)$ is the transition kernel of the Markov chain for $x^{(t)} = x$.

Mykland *et al.* (1995) note that Metropolis–Hastings algorithms are somehow easy to regenerate, given the total freedom in the choice of the transition kernel. First, one can evacuate the atomic part of the kernel by considering only the continuous part, i.e.

$$K(x^{(t)}, y) \geq \rho(x^{(t)}, y) \, q(y|x^{(t)}) = \min \left\{ \frac{f(y)}{f(x^{(t)})} \, q(x^{(t)}|y), q(y|x^{(t)}) \right\} ,$$

where f is the density of the stationary distribution. The determination of s and ν is facilitated in the case of a *pseudo-reversible* transition, i.e. such that there exists a positive function $\tilde{f}$ such that

$$\tilde{f}(x) \, q(y|x) = \tilde{f}(y) \, q(x|y) . \tag{4.3}$$

(Note that $\tilde{f}$ needs not be a probability density and that $q(y|x)$ does not always have a stationary distribution.) In fact, if $w(x) = f(x)/\tilde{f}(x)$, the regeneration parameters are given by

$$s(x) = s_q(x) \, \min \left\{ \frac{c}{w(x)}, 1 \right\} ,$$
$$\nu(x) = \nu_q(x) \, \min \left\{ \frac{w(x)}{c}, 1 \right\} ,$$

for every $c > 0$, where s_q, ν_q are the renewal parameters corresponding to $\tilde{f}$ in (4.2).

In the special case of *independent* Metropolis–Hastings algorithms, $q(y|x) = g(y)$, the pseudo-reversibility condition (4.3) applies with $\tilde{f} = g$. Therefore, $s_q \equiv 1$ and $\nu_q = g$. The lower bound

$$\nu(\xi) \propto g(\xi) \, \min \left\{ \frac{f(\xi)}{cg(\xi)}, 1 \right\}$$
$$\propto \min \left\{ f(\xi), cg(\xi) \right\}$$

behaves like a truncation of the instrumental distribution g depending on the density f to simulate. The regeneration probability (when $y \sim g$ is accepted) is then

$$r(x^{(t)}, \xi) = \begin{cases} \dfrac{c}{w(x^{(t)}) \wedge w(\xi)} & \text{if } w(\xi) \wedge w(x^{(t)}) > c, \\[2ex] \dfrac{w(x^{(t)}) \vee w(\xi)}{c} & \text{if } w(\xi) \vee w(x^{(t)}) < c, \\[2ex] 1 & \text{otherwise,} \end{cases}$$

and c can be selected to increase this probability on the average.

In the case of *symmetric* Metropolis–Hastings algorithms, $q(\xi|x) = q(x|\xi)$ implies that $\tilde{f} \equiv 1$ satisfies the pseudo-reversibility condition (4.3). According to Mykland *et al.* (1995), the parameters s_q and ν_q can be deduced from a set D and a quantity $\tilde{x}$ in the following way:

$$\nu_q(\xi) \propto q(x|\tilde{x}) \, \mathbb{I}_D(\xi) \qquad s_q(x) = \inf_{\xi \in D} \frac{q(\xi|x)}{q(x|\xi)} \,,$$

but the setting is then almost of the same difficulty as the above with the choice of a small set D. Note that $\tilde{x}$ can be calibrated though a preliminary run of the algorithm, using either the mode or the median of $w(x^{(t)})$.

Gilks, Roberts and Sahu (1998) elaborate on the construction of small sets and bounds as in (4.7) by showing that an update of the couple (s, ν) at each renewal event does not jeopardize the ergodic properties of the chain, *even when this construction is based on the whole history of the chain* $(x^{(t)})$, which is a remarkable and most helpful property. In particular, if s is related to a collection of small sets A_i $(i = 1, \ldots, I)$, i.e. is of the form

$$s(x) = \sum_{i=1}^{I} \epsilon_i \mathbb{I}_{A_i}(x),$$

this implies that the collection can be updated after each renewal, by considering the excursions of the chain since the start of the algorithm. More elaborately, if independent or random walk Metropolis–Hastings algorithms are used, the instrumental distributions can be calibrated more carefully, without the usual requirement for a warm-up time after which it must remain unchanged (to become homogeneous).

4.2.3 Splitting the kernel

When (4.1) holds for a triplet (A, ϵ, ν), the transition kernel of the chain $(x^{(t)})$ can be modified without change of stationary distribution. In fact, since

$$
\begin{aligned}
K(x^{(t)}, x^{(t+1)}) &= \epsilon\nu(x^{(t+1)}) + (1 - \epsilon)\frac{K(x^{(t)}, x^{(t+1)}) - \epsilon\nu(x^{(t+1)})}{1 - \epsilon} \\
&= \epsilon\nu(x^{(t+1)}) + (1 - \epsilon)\tilde{K}(x^{(t)}, x^{(t+1)}) \,,
\end{aligned}
$$

where we denote by ν both the bounding measure and its density, and since both terms of the mixture are positive when $x^{(t)} \in A$, we can generate $x^{(t+1)}$ according to

$$x^{(t+1)} = \begin{cases} y_1 \sim \nu(y_1) & \text{with probability } \epsilon, \\ y_2 \sim \tilde{K}(x^{(t)}, y_2) & \text{with probability } 1 - \epsilon, \end{cases} \qquad (4.4)$$

when $x^{(t)} \in A$. The chain is not formally modified since we are still marginally simulating from $K(x^{(t)}, \cdot)$ at each step. However, if we take into account the uniform random variable u_t generated to choose between y_1 and y_2, the decomposition (4.4) introduces independent generations from a distribution ν when $x^{(t)} \in A$ and $u_t < \epsilon$. We then define a sequence of *renewal times* τ_t by $\tau_0 = 1$ and

$$\tau_{t+1} = \inf\{n > \tau_t;\ x^{(n)} \in A \text{ and } u_n \leq \epsilon\}.$$

The blocks $(x^{(\tau_t+1)}, \ldots, x^{(\tau_{t+1})})$ are independent and the partial sums

$$z_t = \sum_{k=\tau_t+1}^{\tau_{t+1}} h(x^{(k)}) \qquad (t \geq 1)$$

are iid. They thus satisfy the following limit theorem under usual regularity conditions:

Lemma 4.2.1 *If* $\mathbb{E}[\tau_1] < \infty$ *and* $h \in L_1(f)$, *the partial sums* z_t *satisfy*

(i) $\quad \displaystyle\sum_{t=1}^{T} z_t / (\tau_{T+1} - \tau_1) \overset{T \to \infty}{\longrightarrow} \mathbb{E}^f[h(x)] \quad (a.s.);$

(ii) $\quad \tau_T / T \overset{T \to \infty}{\longrightarrow} \mathbb{E}^f[\tau_2 - \tau_1] = \mu_A \quad (a.s.).$

Note that, since most MCMC algorithms produce Harris recurrent Markov chains (see Tierney, 1994, and Chan and Geyer, 1994), a finite average return time μ_A to A is usually guaranteed in most cases. Moreover, as also noted by Gilks *et al.*, 1998, this renewal decomposition ensures the applicability of the Central Limit Theorem for the original sum, under the conditions

$$\mathbb{E}[(\tau_{t+1} - \tau_t)^2] < \infty \quad \text{and} \quad \mathbb{E}\left[\left(\sum_{k=1}^{\tau_1} h(x^{(k)})\right)^2\right] < \infty, \qquad (4.5)$$

which imply that the asymptotic variance σ_h^2 is finite (see Meyn and Tweedie, 1993). Indeed, if we denote by $\mu_t = \tau_{t+1} - \tau_t$ the excursion times and by $T(N)$ the number of renewal events before N, the normalized sum

$$\frac{1}{\sqrt{N}} \sum_{n=1}^{N} \left(h(x^{(n)}) - \mathbb{E}^f[h(x)]\right) = \frac{1}{\sqrt{N}} \left\{ \sum_{k=1}^{\tau_1} (h(x^{(k)}) - \mathbb{E}^f[h(x)]) + \right.$$

$$\left. \sum_{t=1}^{T(N)} (z_t - \mu_t \mathbb{E}^f[h(x)]) + \sum_{n=\tau_T+1}^{N} (h(x_n) - \mathbb{E}^f[h(x)]) \right\}$$

is (a.s.) equivalent to

$$\frac{1}{\sqrt{N}} \sum_{t=1}^{T(N)} (z_t - \mu_t \mathbb{E}^f[h(x)])$$

under the conditions (4.5) (since the first and the third terms converge a.s. to 0) and

$$\frac{1}{\sqrt{T(N)}} \sum_{t=1}^{T(N)} (z_t - \mu_t \mathbb{E}^J [h(x)]) \overset{\mathcal{L}}{\longrightarrow} \mathcal{N}(0, \tilde{\sigma}_A^2) \qquad (4.6)$$

by virtue of the usual Central Limit Theorem, the asymptotic variance $\tilde{\sigma}_A^2$ being indexed by the renewal set A. Therefore, the Central Limit Theorem truly applies to the sum of the $(h(x^{(t)}) - \mathbb{E}^J [h(x)])$'s. (See also Gilks *et al.*, 1998, for similar results and Chapter 5 for an extension to Berry-Esséen.)

4.2.4 Splitting in practice

As pointed out in Robert (1995), the modification of the kernel in (4.2) requires simulations from

$$\tilde{K}(x, y) = \frac{K(x, y) - \epsilon \nu(y)}{1 - \epsilon}$$

when $x \in A$, while K is usually unavailable in a closed form. Simulation from $\tilde{K}$ can be achieved by simulating from $K(x, y)$ until acceptance, as follows:

Lemma 4.2.2 *The algorithm*

1. *Simulate* y *from* $K(x,y)$; $[A_{12}]$

2. *Reject* y *with probability* $\epsilon \nu(y)/K(x,y)$.

provides simulations from $\tilde{K}(x,y)$.

This lemma implies the computation of the ratio $\epsilon \nu(y)/K(x,y)$, which is either explicit or which can be approximated by regular Monte Carlo simulations. For instance, the Gibbs sampler kernel can be represented as

$$\begin{aligned} K(x^{(t)}, x^{(t+1)}) &= g_1(x_1 | x_2^{(t)}, \cdots, x_p^{(t)}) \, g_2(x_2 | x_1^{(t+1)}, x_3^{(t)}, \cdots, x_p^{(t)}) \\ &\quad \cdots g_p(x_p | x_1^{(t+1)}, \cdots, x_{p-1}^{(t+1)}), \end{aligned}$$

while, for Data Augmentation,

$$K(x^{(t)}, x^{(t+1)}) = \int g_Z(z|x^{(t)}) g_X(x^{(t+1)}|z) dz$$

can be estimated estimated by

$$\frac{1}{M} \sum_{m=1}^{M} g_X(y|z_m), \qquad (4.7)$$

where the z_m's are iid from $g_Z(z|x^{(t)})$, since (4.7) converges to $K(x^{(t)}, z)$ with M. The examples of §4.5 will illustrate more forcibly this approximation. In Metropolis setups, the transition kernel involves a Dirac mass but the lower bound on K derived from the continuous part of the transition kernel is sufficient to be able to generate from $\tilde{K}(x, y)$.

4.3   Discretization of a continuous Markov chain

Consider now a Markov chain with several disjoint small sets A_i ($i = 1, \ldots, I$) and corresponding parameters (ϵ_i, ν_i). We can define renewal times τ_n ($n \geq 1$) as the successive instants when the Markov chain enters one of these small sets with splitting, i.e. by

$$\tau_n = \inf\{ t > \tau_{n-1};\ \exists\, 1 \leq i \leq I,\ x^{(t-1)} \in A_i \text{ and } x^{(t)} \sim \nu_i \}$$

and $\tau_0 = 1$. (Note that the A_i's ($i = 1, \ldots, I$) are not necessarily a partition of the space.)

From our point of view, a major appeal for this notion of small set is that, although the finite valued sequence deduced from $x^{(t)}$ by

$$\eta^{(t)} = \sum_{i=1}^{I} i\, \mathbb{I}_{A_i}(x^{(t)})$$

is not a Markov chain, the subchain

$$(\xi^{(n)}) = (\eta^{(\tau_n)}),\tag{4.8}$$

i.e. the sequence $(\eta^{(t)})$ sampled only at renewal times, is a Markov chain.

Theorem 4.3.1 *For a Harris recurrent Markov chain $(x^{(t)})$, the sequence $(\xi^{(n)}) = (\eta^{(\tau_n)})$ is an homogeneous Markov chain on the finite state space $\{1, \ldots, I\}$*

Proof. To establish that $(\xi^{(n)})$ is a Markov chain, we need to show that $\xi^{(n)}$ depends on the past only through the last term, $\xi^{(n-1)}$. We have

$$P(\xi^{(n)} = i|\xi^{(n-1)} = j, \xi^{(n-2)} = \ell, \ldots)$$
$$= P\left(\eta^{(\tau_n)} = i|\eta^{(\tau_{n-1})} = j, \eta^{(\tau_{n-2})} = \ell, \ldots\right)$$
$$= P\left(x^{(\tau_n-1)} \in A_i|x^{(\tau_{n-1}-1)} \in A_j, x^{(\tau_{n-2}-1)} \in A_\ell, \ldots\right)$$
$$= \mathbb{E}_{x^{(0)}}\left[\mathbb{I}_{A_i}\left(x^{(\tau_n-1)}\right)\Big|x^{(\tau_{n-1}-1)} \in A_j, x^{(\tau_{n-2}-1)} \in A_\ell, \ldots\right]$$
$$= \mathbb{E}_{x^{(0)}}\left[\mathbb{I}_{A_i}\left(x^{(\tau_{n-1}-1+\Delta_n)}\right)\Big|x^{(\tau_{n-1}-1)} \in A_j, x^{(\tau_{n-2}-1)} \in A_\ell, \ldots\right],$$

where $\Delta_n = \tau_n - \tau_{n-1}$ is independent of $\eta^{(\tau_{n-1})}, \eta^{(\tau_{n-2})}, \ldots$ Therefore, the strong Markov property implies that

$$P(\xi^{(n)} = i | \xi^{(n-1)} = j, \xi^{(n-2)} = \ell, \ldots)$$
$$= \mathbb{E}_{x^{(0)}} \left[\mathbb{I}_{A_i} \left(x^{(\tau_{n-1}-1+\Delta_n)} \right) \Big| x^{(\tau_{n-1}-1)} \in A_j, x^{(\tau_{n-2}-1)} \in A_\ell, \ldots \right]$$
$$= \mathbb{E}_{\nu_j} \left[\mathbb{I}_{A_i} \left(x^{(\Delta_1)} \right) \right]$$
$$= P(\xi^{(n)} = i | \xi^{(n-1)} = j) \,,$$

since $(x^{(t)}, t > \tau_n | x^{(\tau_n)})$ is distributed as $(x^{(t)}, t > \tau_n | \xi^{(n)})$. The homogeneity of the chain can be derived from the invariance (in n) of

$$p_{ji} = P(\xi^{(n)} = i | \xi^{(n-1)} = j),$$

given that $x^{(\tau_{n-1})} \sim \nu_j$ for every n.   □□

Figure 4.1 illustrates discretization on a chain from the Cauchy Benchmark, with three small sets, $B = [-8.5, -7.5]$, $C = [7.5, 8.5]$ and $D = [17.5, 18.5]$, whose construction is explained in §4.5.2. Although the chain visits the three sets quite often, renewal occurs with a much smaller frequency, as shown by the symbols.

FIGURE 4.1. Discretization of a continuous Markov chain from the Cauchy Benchmark, based on three small sets. The renewal events are represented by triangles for B (circles for C and squares for D, respectively). (*Source:* Guihenneuc–Jouyaux and Robert, 1998.)

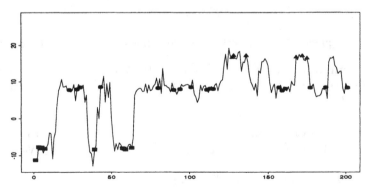

This result is fundamental to our overall purpose of controlling Markov chains through their discretized counterparts, since Theorem 4.3.1 shows that finite Markov chains can be rigorously derived from a continuous Markov chain in most setups. The drawback is obviously that the small sets need to be exhibited, but, as shown by Mykland *et al.* (1995), quasi-automatic schemes are available. Moreover, the choice of the space where the small sets are constructed is open and, at least for Gibbs samplers, there are often obvious choices when the Duality Principle of Chapter 1 applies.

4.4 Convergence assessment through the divergence criterion

Once a finite state space chain is obtained, the whole range of finite Markov chain theory is available, providing a variety of different convergence results whose conjunction can strengthen the convergence diagnosis. For instance, Raftery and Lewis (1996) use a normal approximation for the average number of visits to a given state. We choose instead to use an exact evaluation of the mixing rate of the chain based on the comparison between the number of visits to a given state from two different starting points. This *divergence* evaluation is derived from Kemeny and Snell (1960) and, besides its simplicity and elegance, it meets the main requirement of convergence control since it compares the behaviour of chains with different starting points till independence. See Chapter 5 for another convergence criterion based on the Central Limit Theorem.

4.4.1   The divergence criterion

In the study of regular Markov chains with transition matrix $\mathbb{P}$, Kemeny and Snell (1960) point out the importance of the so-called *fundamental matrix*

$$\mathbb{Z} = [\mathbf{I} - (\mathbb{P} - \mathbf{A})]^{-1},$$

where $\mathbf{A}$ is the limiting matrix $\mathbb{P}^\infty$, with all rows equal to π, the stationary distribution associated with $\mathbb{P}$. A particular property of interest is that, if $N_j(T)$ denotes the number of times the Markov chain $(x^{(t)})$ is in state j $(1 \leq j \leq I)$ in the first T stages, i.e.

$$N_j(T) = \sum_{t=1}^{T} \mathbb{I}_j(x^{(t)}),$$

then, for any initial distribution π_0, the so-called *divergence*

$$\operatorname{div}_j(\pi_0, \pi) = \lim_{T \to \infty} \mathbb{E}_{\pi_0}[N_j(T)] - T\pi_j \qquad (4.9)$$

satisfies

$$\operatorname{div}_j(\pi_0, \pi) = \sum_{\ell=1}^{I} \pi_{0\ell} z_{\ell j} - \pi_j \qquad (4.10)$$

(see Kemeny and Snell, 1960, for a proof). A consequence of (4.10) is that, for two arbitrary initial distributions π_0 and π_0',

$$
\mathrm{div}_j(\pi_0, \pi_0') = \lim_{T \to \infty} \left\{ \mathbb{E}_{\pi_0} \left[\sum_{t=1}^{T} \mathbb{I}_j(x^{(t)}) \right] - \mathbb{E}_{\pi_0'} \left[\sum_{t=1}^{T} \mathbb{I}_j(x^{(t)}) \right] \right\}
$$

$$
= \sum_{\ell=1}^{I} (\pi_{0\ell} - \pi_{0\ell}') z_{\ell j} .
$$

In particular, if two chains start from states u and v with corresponding numbers of passages in j, $N_j^u(T)$ and $N_j^v(T)$ respectively, the limiting difference is

$$
\mathrm{div}_j(u, v) = \lim_{T \to \infty} \left\{ \mathbb{E}_u \left[\sum_{t=1}^{T} \mathbb{I}_j(x^{(t)}) \right] - \mathbb{E}_v \left[\sum_{t=1}^{T} \mathbb{I}_j(x^{(t)}) \right] \right\}
$$

$$
= z_{uj} - z_{vj} . \tag{4.11}
$$

The relevance of this notion for convergence control purposes is multiple. First, it assess the effect of initial values on the chain by exhibiting the right scale for the convergence of (4.9) to a finite limit. Indeed, each term

$$
\mathbb{E}_u \left[\sum_{t=1}^{\infty} \mathbb{I}_j(\theta^{(t)}) \right]
$$

is infinite since the chains are recurrent. In that sense, the convergence result (4.11) is stronger that the Ergodic Theorem, since the later only indicates independence from initial values in the scale $1/T$. Note the similarity of

$$
\sum_{t=1}^{T} \mathbb{I}_j(\theta^{(t)}) - T\pi_j
$$

with Yu and Mykland's (1998) *CUSUM's criterion* (see §2.2.1), the difference being that π_j is estimated from the same chain in their case. Moreover, the criterion is immediately valid in the sense that it does not require stationarity but, on the opposite, takes into account the initial values. A third incentive is that the property that the limiting difference in the number of visits is equal to $(z_{uj} - z_{vj})$ provides a quantitative control tool, which is that the matrix $\mathbb{Z}$ can be estimated directly from the transitions of the Markov chain. We thus obtain a *control variate* technique for general Markov chains since the estimates of $\mathrm{div}_j(u, v)$ and of $(z_{uj} - z_{vj})$ must converge to the same quantity.

4.4.2   A finite example

While the matrix $\mathbb{Z}$ is usually unknown, (4.11) is still of interest for convergence control purposes. More specifically, a first implementation of the method is to generate in parallel $M \times I$ chains starting from the I different states, the M replications being used to evaluate the expectations $\mathbb{E}_u[N_j^u(T)]$ under each possible starting state u by simple averages on the parallel chains. The convergence criterion is then based on the stabilization (in T) of the approximation of $\mathbb{E}_u[N_j^u(T)] - \mathbb{E}_v[N_j^v(T)]$

$$\frac{1}{M} \sum_{m=1}^{M} (N_j^{u,m}(T) - N_j^{v,m}(T)) = \frac{1}{M} \sum_{m=1}^{M} \sum_{t=1}^{T} \left\{ \mathbb{I}_j(x_{u,m}^{(t)}) - \mathbb{I}_j(x_{v,m}^{(t)}) \right\},$$

(4.12)

for all triplets (j, u, v), where $N_j^{u,m}(t)$ denotes the number of passages in state j before time t for the m-th replication of the chain $(x_{u,m}^{(t)})$ starting from state u. However, as shown below, this approach exhibits severe instability since it requires huge values of M to converge.

Consider a three state Markov chain with transition matrix

$$\mathbb{P} = \begin{pmatrix} 0.07 & 0.32 & 0.61 \\ 0.41 & 0.11 & 0.48 \\ 0.63 & 0.21 & 0.16 \end{pmatrix}.$$

For different values of M ($M = 100, 1000, 10,000$), we compute the difference (4.12) when the chains starting in states 0, 1 and 2 are run independently, up to $T = 10,000$. Figure 4.2 shows that stability is not attained, even if the passage to $10,000$ parallel chains reduces the variations of (4.12) (note that the scales are different).

Note that an additional convergence assessment is to compare the current approximation of $\text{div}_j(u, v)$ with the known or estimated limit $z_{uj} - z_{vj}$. In this particular case, the limits for triplets $(1, 0, 1)$ and $(2, 0, 2)$ are -0.833 and -0.695. The reason for the lack of stability of the indicator (4.12) is the very erratic behaviour of the r.v. $N_j^u(T)$ rather than the slow convergence (in T) of the difference $\mathbb{E}[N_j^u(T) - N_j^v(T)]$ to (4.11), since the graphs fluctuate around the limiting value with crossings even for small values of T. Note that this phenomenon is not surprising from a probabilistic point of view, since it reproduces a coin tossing experiment as in Feller (1970, Chapter 3), who shows that long excursions from the average are the rule.

4.4.3 Stopping rules

A superior alternative in the implementation of the divergence criterion is to use *stopping rules* to accelerate convergence. In fact, when two chains have met in an arbitrary state, their paths follow the same distribution from this meeting time and the additional terms $\mathbb{I}_j(x_u^{(t)}) - \mathbb{I}_j(x_v^{(t)})$ in (4.12) are merely noise. It therefore makes sense to stop the evaluation in (4.12) at this meeting time, in a spirit similar to *coupling* (see §2.4). The computation of

$$\lim_{T \to \infty} \left\{ \mathbb{E}[N_j^u(T)] - \mathbb{E}[N_j^v(T)] \right\}$$

FIGURE 4.2. Estimation of the difference (4.11) for two triplets (j, u, v) and different M's. Full lines stand for independent approximations (4.12) and dashed lines for limiting values, -0.833 and -0.695.

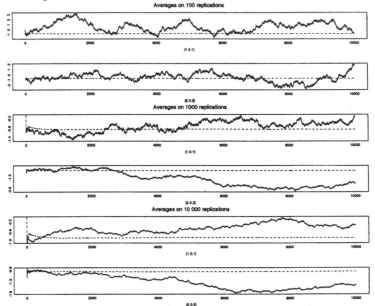

is therefore greatly simplified since the limit is not relevant anymore. More formally, if $T(u, v)$ denotes the stopping time when the—independent or coupled—chains $(x_u^{(t)})$ and $(x_v^{(t)})$, starting from states u and v respectively, are identical for the first time, we have the following result:

Lemma 4.4.1 *For every coupling such that* $\mathbb{E}[T(u, v)^2] < \infty$, *the divergence* $\mathrm{div}_j(u, v)$ *is given by*

$$\mathbb{E}\left[\sum_{t=1}^{T(u,v)} \left\{\mathbb{I}_j(x_u^{(t)}) - \mathbb{I}_j(x_v^{(t)})\right\}\right].$$

Proof. Since $T(u, v)$ is a stopping time, the strong Markov property implies that, for $n > \ell$,

$$\mathbb{E}\left[h(\xi_u^{(n)}) \mid \xi_u^{(T(u,v))} = j, T(u, v) = \ell\right] = \mathbb{E}_j\left[h(\xi^{(n-\ell)})\right]$$

$$= \mathbb{E}\left[h(\xi_v^{(n)}) \mid \xi_v^{(T(u,v))} = j, T(u, v) = \ell\right]$$

for every function h and, by conditioning, we derive that

$$\mathbb{E}\left[\sum_{t=1}^{T} \left\{\mathbb{I}_j(\xi_u^{(t)}) - \mathbb{I}_j(\xi_v^{(t)})\right\}\right]$$

$$= \mathbb{E}\left[\sum_{t=1}^{T \wedge T(u,v)} \left\{ \mathbb{I}_j(\xi_u^{(t)}) - \mathbb{I}_j(\xi_v^{(t)}) \right\} \right]$$

$$= \mathbb{E}\left[\sum_{t=1}^{T(u,v)} \left\{ \mathbb{I}_j(\xi_u^{(t)}) - \mathbb{I}_j(\xi_v^{(t)}) \right\} \mathbb{I}_{T(u,v) \le T} \right]$$

$$+ \mathbb{E}\left[\sum_{t=1}^{T} \left\{ \mathbb{I}_j(\xi_u^{(t)}) - \mathbb{I}_j(\xi_v^{(t)}) \right\} \mathbb{I}_{T(u,v) > T} \right]$$

Now,

$$\left| \mathbb{E}\left[\sum_{t=1}^{T} \left\{ \mathbb{I}_j(\xi_u^{(t)}) - \mathbb{I}_j(\xi_v^{(t)}) \right\} \mathbb{I}_{T(u,v) > T} \right] \right| \le TP(T(u,v) > T)$$

$$\le \mathbb{E}[T(u,v)^2]/T$$

implies that the left hand side goes to 0 when T goes to infinity. And

$$\mathbb{E}\left[\sum_{t=1}^{T(u,v)} \left\{ \mathbb{I}_j(\xi_u^{(t)}) - \mathbb{I}_j(\xi_v^{(t)}) \right\} \right]$$

$$- \mathbb{E}\left[\sum_{t=1}^{T(u,v)} \left\{ \mathbb{I}_j(\xi_u^{(t)}) - \mathbb{I}_j(\xi_v^{(t)}) \right\} \mathbb{I}_{T(u,v) \le T} \right]$$

$$= \mathbb{E}\left[\sum_{t=1}^{T(u,v)} \left\{ \mathbb{I}_j(\xi_u^{(t)}) - \mathbb{I}_j(\xi_v^{(t)}) \right\} \mathbb{I}_{T(u,v) > T} \right]$$

$$\le 2\mathbb{E}\left[T(u,v) \mathbb{I}_{T(u,v) > T} \right]$$

goes to 0 when T goes to infinity by the Dominated Convergence Theorem. Therefore,

$$\lim_{T \to \infty} \mathbb{E}\left[\sum_{t=1}^{T} \left\{ \mathbb{I}_j(\xi_u^{(t)}) - \mathbb{I}_j(\xi_v^{(t)}) \right\} \right] = \mathbb{E}\left[\sum_{t=1}^{T(u,v)} \left\{ \mathbb{I}_j(\xi_u^{(t)}) - \mathbb{I}_j(\xi_v^{(t)}) \right\} \right].$$

□□

The condition $\mathbb{E}[T(u,v)^2] < \infty$ holds in the cases when the parallel chains are independent and when they are coupled in a deterministic fashion (see §2.4.1), namely when all parallel chains are based on the same sequence of uniforms. In fact, both setups can be rewritten in terms of a single Markov chain and $T(u,v)$ is then the time necessary to reach a certain union of states. It is thus rarely necessary to verify that $\mathbb{E}[T(u,v)^2] < \infty$

holds in practice. (Note also that this condition can be replaced by the weaker (but more formal) condition $\mathbb{E}[T(u,v)^{1+\epsilon}] < \infty$ for an arbitrary $\epsilon > 0$.)

In practice, if $T_m(u,v)$ is the first time when $(x_{u,m}^{(t)})$ and $(x_{v,m}^{(t)})$ are equal, the divergence $\mathrm{div}_j(u,v)$ can be approximated by

$$\frac{1}{M} \sum_{m=1}^{M} \left\{ N_j^{u,m}[T_m(u,v)] - N_j^{v,m}[T_m(u,v)] \right\}$$

$$= \frac{1}{M} \sum_{m=1}^{M} \sum_{t=1}^{N_m(u,v)} \left[\mathbb{I}_j(x_{u,m}^{(t)}) - \mathbb{I}_j(x_{v,m}^{(t)}) \right]. \qquad (4.13)$$

As mentioned above, the I chains $(x_{u,m}^{(t)})$ ($1 \le u \le I$) can be coupled in a deterministic manner by using the same underlying uniform variables $(u_m^{(t)})$, namely by generating $x_{u,m}^{(t)}$ by

$$x_{u,m}^{(t)} = i \qquad \text{if, and only if,} \qquad \sum_{r=1}^{i-1} p_{jr} < u_m^{(t)} \le \sum_{r=1}^{i} p_{jr}$$

when $x_{u,m}^{(t-1)} = j$. The improvement brought by coupling in the evaluation of (4.11) is quite significant since the dotted lines in Figure 4.2 actually correspond to the approximation (4.13). Quite markedly, stability for the coupled divergences occurs in less than 1000 iterations. It could be argued that the comparison is biased since each new value for the coupled divergence graph relies on a stopping rule, which is that the three Markov chains have all met, thus involves a random number of terms, but the average time till coupling is 2.62 (to be compared with M).

4.4.4   Extension to continuous state spaces

The implementation of the divergence control method in the continuous case is quite similar to the above proposal. For a given replication m of the M parallel runs necessary to evaluate the expectation (4.11), I chains $(x_j^{(t)})$ are initialized from the I bounding measures ν_j ($1 \le j \le I$). The generation of $x_j^{(t)}$ is modified according to (4.4) when $x_j^{(t-1)}$ enters one of the small sets A_i and this modification provides the subchains $(\xi_j^{(n)})$. The contribution of the m-th replication to the approximation of $\mathrm{div}_j(i_1, i_2)$, namely

$$\lim_{N \to \infty} \sum_{n=1}^{N} \left[\mathbb{I}_j(\xi_{i_1}^{(n)}) - \mathbb{I}_j(\xi_{i_2}^{(n)}) \right],$$

is actually given by

$$\sum_{n=1}^{T(i_1,i_2)} \left\{ \mathbb{I}_j(\xi_{i_1}^{(n)}) - \mathbb{I}_j(\xi_{i_2}^{(n)}) \right\} , \qquad (4.14)$$

where $T(i_1, i_2)$ is the stopping time corresponding to the first occurrence of $\xi_{i_1}^{(n)} = \xi_{i_2}^{(n)}$, since (4.14) is an unbiased estimator of $\text{div}_j(i_1, i_2)$ according to Lemma 4.4.1.

A first convergence assessment is derived from the graphical evaluation of the stabilization of the estimated divergences. It can be improved by the additional estimation of the variance of (4.14), since the parallel runs (in m) are independent (see Figures 4.4–4.6). A more advanced convergence assessment follows from the comparison of the estimated divergences with the estimated limits $z_{i_1 j} - z_{i_2 j}$, when the transition matrix of $(\xi^{(n)})$ is derived from the various chains and the fundamental matrix $\mathbb{Z}$ is computed with this approximation.

Deterministic coupling cannot be easily implemented in continuous state space chains (see §2.4), since generations from continuous distributions usually require a random number of uniform r.v.'s (see Devroye, 1985, or Robert and Casella, 1998). However, we are only interested in the chain $(\xi^{(n)})$ and can thus create coupling at this discrete stage. In fact, two departures from independence on the parallel chains can accelerate convergence for the approximation of (4.11). First, the same uniform r.v. can be used at each (absolute) time t to decide whether this is a renewal time for every chain entering an arbitrary small set A_j. Second, traditional antithetic arguments can be transferred to this setting in order to accelerate a meeting in the same small set. As shown below, it is also possible to use common uniforms for the parallel chains at different stages if some group structure applies.

4.4.5 From divergence estimation to convergence control

When considering the approximation of the divergence factors and, more generally, the convergence of the chain, there is now a well-documented literature about the dangers of using solely parallel chains (see §2.3 and Geyer, 1992, Raftery and Lewis, 1996), because of the dependence on starting values and of the waste of simulations. One can argue that the present setup is different because of the renewal mechanism, which somehow eliminates the dependence on the starting values. For instance, there is no objection to estimate the transition matrix $\mathbb{P}$ of the finite chain $(\xi^{(n)})$ in (4.8) from these parallel chains. But, while earlier criticisms still apply in this case (like the waste of simulations or the impossibility of getting an assessment "on line"), the call for parallel chains has rather different negative features in the present setup. Indeed, the sample produced by the final values of the parallel chains cannot be exploited as a stationary sample of the distribution of interest because of the short runs created by the stopping rule. Moreover, the fact of starting an equal number of chains from each small set does not necessarily reflect the weights of these sets in the stationarity distribution. In that sense, the method is the opposite of an 'on-line' control technique, even though it provides useful information on the mixing rate of the chain.

We now show how the divergence criterion can be implemented by using *only two* parallel chains, whatever the numbers of small sets and of triplets (ℓ, i_1, i_2). This alternative implementation is based on *Birkhoff's pointwise ergodic theorem* (see, e.g., Battacharya and Waymire, 1990, pp. 223-227, for a proof), which extends the standard ergodic theorem (see Meyn and Tweedie, 1993) to functionals of the chain directed by a stopping time. We denote $\mathbf{X} = (x^{(1)}, \ldots)$ a Markov chain and $D^r \mathbf{X} = (x^{(r+1)}, x^{(r+2)}, \ldots)$ the r-shifted version of the chain.

Theorem 4.4.2 (Birkhoff) *For an ergodic Markov chain $(x^{(t)})$, with stationary distribution f, and a functional g of $\mathbf{X}$, the average*

$$\frac{1}{M} \sum_{m=1}^{M} g(D^m \mathbf{X}) \tag{4.15}$$

converges almost surely to the expectation $\mathbb{E}^f(\mathbf{X})]$.

In particular, if S is a *stopping time* (see Meyn and Tweedie, 1993, p.71) and if g satisfies

$$g(\mathbf{X}) = g(x^{(1)}, \ldots, x^{(S(\mathbf{X}))}),$$

the above result applies. This theorem thus implies that overlapping parts of a Markov chain can contribute to the approximation of $\mathbb{E}^f[g(\xi^{(1)}, \ldots)]$ without being detrimental to the convergence of the standard average. For instance, if $\{S = T\}$ is only function of the value of the chain in T, $\xi^{(T)}$, as it is in our case, Birkhoff's ergodic theorem allows us to use $(\xi^{(1)}, \ldots, \xi^{(T)})$, then $(\xi^{(2)}, \ldots, \xi^{(T)})$, etc., up to $(\xi^{(T-1)}, \xi^{(T)})$ and $\xi^{(T)}$ as if $\xi^{(T)}$ was not repeated T times in these samples.

In the setup of this paper, Birkhoff's ergodic theorem can be invoked to use only two chains $(x_1^{(t)})$ and $(x_2^{(t)})$ with arbitrary starting points u and v since, for every n, $(\xi_1^{(n)}, \xi_2^{(n)})$ can contribute to the evaluation of the corresponding $\mathrm{div}_\ell(\xi_1^{(n)}, \xi_2^{(n)})$ in the usual manner, namely by counting out the difference in the numbers of visits to state ℓ between n and the next meeting time of $\xi_1^{(t)}$ and $\xi_2^{(t)}$. In fact, if ℓ in $\mathrm{div}_\ell(i_1, i_2)$ is a specific function of (i_1, i_2), the function g can be constructed as a vector of the approximations (4.13) of the $\mathrm{div}_j(i_1, i_2)$'s, which involves a stopping rule.

The gain brought by this result is far from negligible since, instead of using a couple of (independent or not) chains $(\xi_1^{(n)}, \xi_2^{(n)})$ only once between the starting point and their stopping time N, the same sequence is used several times in the sum (4.15) and contributes to the estimation of the divergences for the values $(\xi_1^{(m)}, \xi_2^{(m)}) = (u, v)$ $(m = 1, \ldots, n)$. Moreover, the method no longer requires to re-start the chains $\xi_1^{(n)}$ and $\xi_2^{(n)}$ once they have met and this feature allows for an on-line control of the MCMC algorithm, a better mixing of the chain and a direct use for estimation purposes. In fact, the continuous chains $(\theta_i^{(n)})$ behind the discretized subchains $(\xi_i^{(n)})$ $(i = 1, 2)$ are generated without any constraint and the resulting $(\xi_1^{(n)}, \xi_2^{(n)})$'s are used to update the divergence estimations by batches, that is every time $\xi_1^{(n)} = \xi_2^{(n)}$.

4.5   Illustration for the benchmark examples

We now consider the examples introduced in Chapter 1 in order to show how our technique applies and performs. Note that, although all examples involve data augmentation algorithms where the derivation of the small set is usually straightforward, the minorizing techniques can be easily extended to other MCMC setups, including Metropolis–Hastings algorithms.

4.5.1   Pump Benchmark

The nuclear pump failure dataset of Gaver and O'Muircheartaigh (1987) has been introduced in Example 1.6 and the data is described in Table 1.2. As shown in §1.3, the space $(\mathbb{R}_+)$ is a small set and (β^t) is uniformly ergodic but the probability of renewal on $(\mathbb{R}_+)$

$$\epsilon \propto \frac{\delta^{\gamma+10\alpha}}{\Gamma(10\alpha + \gamma)},$$

is much too small to be used for convergence control (see Robert, 1996c, Chapter 6).

If we now introduce the small sets $A_j = [\underline{\beta}_j, \overline{\beta}_j]$ $(j = 1, \ldots, J)$, the lower bound on the transition kernel is

$$
\begin{aligned}
K(\beta, \beta') \geq & \int \frac{(\beta')^{\gamma+10\alpha-1}}{\Gamma(10\alpha+\gamma)} \left(\delta + \sum_{i=1}^{10} \lambda_i\right)^{\gamma+10\alpha} e^{-\beta'\left(\delta+\sum_{i=1}^{10} \lambda_i\right)} \\
& \times \prod_{i=1}^{10} \left\{ \frac{(t_i + \underline{\beta}_j)^{p_i+\alpha}}{\Gamma(p_i + \alpha)} \lambda_i^{p_i+\alpha-1} e^{-(t_i+\overline{\beta}_j)\lambda_i} \right\} d\lambda_1 \cdots d\lambda_{10} \\
= & \prod_{i=1}^{10} \left\{ \frac{(t_i + \underline{\beta}_j)^{p_i+\alpha}}{(t_i + \overline{\beta}_j)^{p_i+\alpha}} \right\} \int \frac{(\beta')^{\gamma+10\alpha-1}}{\Gamma(10\alpha+\gamma)} \left(\delta + \sum_{i=1}^{10} \lambda_i\right)^{\gamma+10\alpha} \\
& \times e^{-\beta'\left(\delta+\sum_{i=1}^{10} \lambda_i\right)}
\end{aligned}
$$

$$\times \prod_{i=1}^{10} \left\{ \frac{(t_i + \overline{\beta}_j)^{p_i+\alpha}}{\Gamma(p_i + \alpha)} \, \lambda_i^{p_i+\alpha-1} \, e^{-(t_i+\overline{\beta}_j)\lambda_i} \right\} d\lambda_1 \cdots d\lambda_{10} \, .$$

The probability of renewal within a small set A_j is therefore $(j = 1, \dots, J)$

$$\epsilon_j = \prod_{i=1}^{10} \frac{(t_i + \underline{\beta}_j)^{p_i+\alpha}}{(t_i + \overline{\beta}_j)^{p_i+\alpha}} \, ,$$

while the bounding probability ν_j is the marginal distribution (in β) of the joint distribution

$$\lambda_i \sim \mathcal{G}a(p_i + \alpha, t_i + \overline{\beta}_j), \qquad \beta | \lambda_1, \dots, \lambda_{10} \sim \mathcal{G}a\left(\gamma + 10\alpha, \delta + \sum_{i=1}^{10} \lambda_i\right).$$

A preliminary run of the Gibbs sampler on 5000 iterations provides the small sets given in Table 4.1 as those maximizing the probability of renewal $\varrho_j = \epsilon_j P(\beta^{(t)} \in A_j)$ $(j = 1, \dots, 8)$. As shown by Figure 4.3, they are concentrated in the center of the posterior distribution of β.

TABLE 4.1. Small sets associated with the transition kernel $K(\beta, \beta')$ and corresponding renewal parameters $(j = 1, \dots, 8)$.

A_j	$[1.6, 1.78]$	$[1.8, 1.94]$	$[1.95, 2.13]$	$[2.15, 2.37]$
ϵ_j	0.268	0.372	0.291	0.234
ϱ_j	0.0202	0.0276	0.0287	0.0314

A_j	$[2.4, 2.54]$	$[2.55, 2.69]$	$[2.7, 2.86]$	$[2.9, 3.04]$
ϵ_j	0.409	0.417	0.377	0.435
ϱ_j	0.0342	0.0299	0.0258	0.0212

Following the developments above, the convergence assessment associated with these small sets A_j can be based on parallel runs of eight chains $(\beta_j^{(t)})$ $(j = 1, \dots, 8)$ starting from the eight small sets with initial distributions the corresponding ν_j:

1. Generate $(i = 1, \dots, 8)$

$$\lambda_i \sim \mathcal{E}xp(t_i + \overline{\beta}_j);$$

2. Generate   [A_{13}]

$$\beta \sim \mathcal{G}a\left(\gamma + 10\alpha, \delta + \sum_{i=1}^{8} \lambda_i\right).$$

FIGURE 4.3. Posterior distribution of β for the Pump Benchmark obtained by Gibbs sampling (5000 iterations). (*Source:* Guihenneuc–Jouyaux and Robert, 1998.)

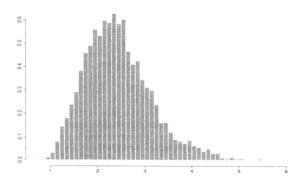

The chains $(\beta_j^{(t)})$ induce corresponding finite state space chains $(\xi_j^{(n)})$ with $\xi_j^{(1)} = j$ and contribute to the approximation of the divergences $\mathrm{div}_\ell(i_1, i_2)$ via the sums (4.14), depending on coupling times $N(i_1, i_2)$. Figure 4.4 describes the convergence of four estimated divergences as the number of parallel runs increases. For each couple (i_1, i_2), the corresponding state ℓ is $i_2(\mathrm{mod}\ 8) + 1$. The average stabilize rather quickly and, moreover, the overall number of iterations required by the method is moderate since the mean coupling time is only 14.0; this implies that each sum (4.14) involves on average 14 steps of the Gibbs sampler. The standard deviation is derived from the empirical variance of the sums (4.14).

In order to approximate the ratio $\epsilon \nu(\beta')/K(\beta, \beta')$ mentioned in §3.2, the integrals in both $\nu(\beta')$ and $K(\beta, \beta')$ are replaced by sums, leading to the approximation

$$\frac{\epsilon \nu(\beta')}{K(\beta, \beta')} \simeq \frac{\sum_{s=1}^{S} \left(\delta + \sum_{i=1}^{10} \lambda_i^s\right)^{\gamma + 10\alpha} \exp\left\{-\beta' \sum_{i=1}^{10} \lambda_i^s\right\}}{\sum_{s=1}^{S} \left(\delta + \sum_{i=1}^{10} \tilde{\lambda}_i^s\right)^{\gamma + 10\alpha} \exp\left\{-\beta' \sum_{i=1}^{10} \tilde{\lambda}_i^s\right\}},$$

where the λ_i^s are generated from $\mathcal{E}xp(t_i + \overline{\beta}_j)$ and the $\tilde{\lambda}_i^s$ from $\mathcal{E}xp(t_i + \beta)$. This approximation device is theoretically justified for S large enough although it increases the computational time. An accelerating (and stabilizing) technique is to use repeatedly the same sample of S exponential $\mathcal{E}xp(1)$ r.v.'s for the generation of the $(t_i + \beta)\tilde{\lambda}_i^s$'s, that is, to take advantage of the scale structure of the gamma distribution. In the simulations, we took $S = 500$, although smaller values of S ensure stability of the approximation.

FIGURE 4.4. Convergence of the divergence criterion based on (4.11) for four chains started from four small sets A_j in the Pump Benchmark. The triplets (i_1, i_2, ℓ) index the difference in the number of visits of ℓ by the chains $(\xi_{i_1}^{(t)})$ and $(\xi_{i_2}^{(t)})$. The envelope is located two standard deviations from the average. For each replication, the chains are restarted from the corresponding small set. The theoretical limits derived from the estimation of $\mathbb{P}$ are $-.00498$, $-.0403$, $.00332$ and $-.00198$ (based on $50,000$ iterations). (*Source:* Guihenneuc–Jouyaux and Robert, 1998.)

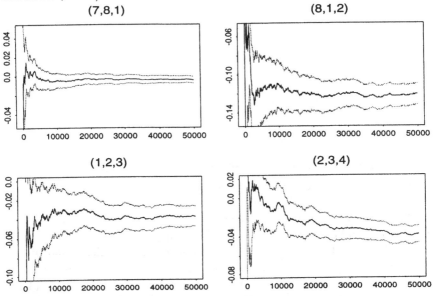

4.5.2   Cauchy Benchmark

Consider the posterior distribution (1.6) of Example 1.2.3 and the associated Gibbs sampler in Example 2.2.1. If we take $A = [r_1, r_2]$ with $x_2 \in [r_1, r_2]$, $x_1 < r_1$ and $x_3 > r_2$ (assuming $x_1 < x_2 < x_3$), the following bounds

$$
\begin{aligned}
\rho_{11} = r_1 - x_1 < |\theta - x_1| \;&<\; \rho_{12} = r_2 - x_1, \\
0 < |\theta - x_2| \;&<\; \rho_{22} = \max(r_2 - x_2, x_2 - r_1), \\
\rho_{31} = x_3 - r_2 < |\theta - x_3| \;&<\; \rho_{32} = x_3 - r_1
\end{aligned}
$$

induce a minorizing probability measure ν and a corresponding constant ϵ. Indeed,

$$
K(\theta, \theta') \;\geq\; \int_{\mathbb{R}_+^3} \exp\left\{ -\left(\theta' - \tau^2(\eta_1 x_1 + \eta_2 x_2 + \eta_3 x_3)\right)^2 / 2\tau^2 \right\}
$$

$$\times \tau \frac{1}{\sqrt{2\pi}} \frac{1 + \rho_{11}^2}{2} \exp\{-(1 + \rho_{12}^2)\eta_1/2\}$$

$$\times \frac{1}{2} \exp\{-(1 + \rho_{22}^2)\eta_2/2\} \frac{1 + \rho_{31}^2}{2}$$

$$\times \exp\{-(1 + \rho_{32}^2)\eta_3/2\} \, d\eta_1 d\eta_2 d\eta_3$$

$$= \frac{1 + \rho_{11}^2}{1 + \rho_{12}^2} \frac{1}{1 + \rho_{22}^2} \frac{1 + \rho_{31}^2}{1 + \rho_{32}^2} \nu(\theta') = \epsilon\nu(\theta'),$$

where ν is the density of the marginal distribution (in θ) of

$$(\theta, \eta_1, \eta_2, \eta_3) \sim \mathcal{N}\left(\tau^2(\eta_1 x_1 + \eta_2 x_2 + \eta_3 x_3), \tau^2\right)$$

$$\times \mathcal{E}xp\left(\frac{1 + \rho_{12}^2}{2}\right) \mathcal{E}xp\left(\frac{1 + \rho_{22}^2}{2}\right) \mathcal{E}xp\left(\frac{1 + \rho_{32}^2}{2}\right).$$

Similar derivations can be obtained for the sets $B = [s_1, s_2]$ with $s_1 < x_1 < s_2 < x_2$ and $D = [v_1, v_2]$ with $x_2 < v_1 < x_3 < v_3$. If we choose in addition $s_2 < r_1$ and $r_2 < v_1$, the three small sets are disjoint and we can thus create a three-state Markov chain. A preliminary run of the Gibbs sampler (1.6) on 5000 iterations leads to the choice of the three small sets

$$B = [-8.5, -7.5], \quad C = [7.5, 8.5], \quad D = [17.5, 18.5]$$

as optimizing the probabilities of renewal,

$$\varrho_B = 0.02, \qquad \varrho_C = 0.10, \qquad \varrho_D = 0.05.$$

Figure 4.1 gives the 200 first values of $\theta^{(t)}$ generated from (1.6) and indicates the corresponding occurrences of $\xi^{(n)}$. Note that the choice of the small sets is by no means restricted to neighbourhoods of the modes, although this increases the probability of renewal.

As above, the convergence assessment is based on the replication of three parallel chains and coupling. Figure 4.5 describes the convergence of the average of the sums (1.6) for the three triplets (B, C, D), (C, D, B) and (D, B, C) in the setup of the Cauchy benchmark. The average coupling time in this case is 38. Therefore, the simulation of $M = 50,000$ parallel chains corresponds to nearly two million iterations of the Gibbs sampler.[1] Additional coupling links between the chains are created by taking advantage of the scale structures of the proposal distributions to use the same samples from $\mathcal{E}xp(1)$ and $\mathcal{N}(0, 1)$ whenever possible.

If we now consider the alternative implementation following from the use of Birkhoff's ergodic theorem for stopping times, Figure 4.6 illustrates the convergence of the divergence control device, the starting small sets

---

[1] Chapter 5 will show that this model does not require such a large number of iterations to achieve approximate normality.

FIGURE 4.5. Convergence of the divergence criterion based on (4.11) for three chains started from the three small sets B, C and D in the Cauchy benchmark. The triplets (ℓ, i_1, i_2) index the difference in the number of visits of ℓ by the chains $\xi_{i_1}^{(m)}$ and $\xi_{i_2}^{(m)}$. The envelope is located two standard deviations from the average. When the three couples have met, the three chains are restarted from the three small sets. (*Source:* Guihenneuc–Jouyaux and Robert, 1998.)

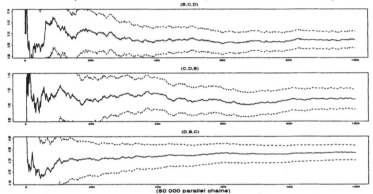

FIGURE 4.6. Convergence of the divergence criterion for two chains initially started from B and D in the Cauchy benchmark. The triplets (ℓ, i_1, i_2) index the difference in the number of visits of ℓ by chains starting from i_1 and from i_2. The envelope is located two standard deviations from the average. The theoretical limits derived from the estimation of $\mathbb{P}$ are 0.094, 1.202 and -0.688. The scales of the three graphs represent the number of times $\mathrm{div}_\ell(i_1, i_2)$ has been updated along the 500,000 iterations. The final values of the three estimated divergences are 0.164, 1.076 and -0.825. (*Source:* Guihenneuc–Jouyaux and Robert, 1998.)

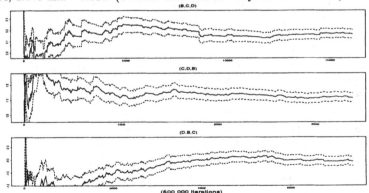

being B and C. It is comparable to the above fully parallel implementation for a number of simulations of $(\theta^{(t)})$ which is much smaller (since 500,000 iterations in this case is to be compared with 2 million iterations in the above setup).

4.5.3 Multinomial Benchmark

Since the Duality Principle applies in this setup (see Chapter 1), there already exists a finite state space Markov chain which controls the convergence of the Markov chain of interest. For comparison purposes, we can still derive a small set on the (μ, η) space. Besides, this example has the additional appeal to be multidimensional in the chain of interest, a feature which is supposed to slow down convergence assessment through renewal methods, according to Gilks, Roberts and Sahu (1998).

Consider thus the transition kernel on the (μ, η) chain:

$$
K((\mu, \eta), (\mu', \eta')) = \sum_{(z_1, z_2, z_3, z_4)} \binom{x_1}{z_1} \left(\frac{a_1 \mu}{a_1 \mu + b_1} \right)^{z_1} \left(\frac{b_1}{a_1 \mu + b_1} \right)^{x_1 - z_1}
$$

$$
\times \binom{x_2}{z_2} \left(\frac{a_2 \mu}{a_2 \mu + b_2} \right)^{z_2} \left(\frac{b_2}{a_2 \mu + b_2} \right)^{x_2 - z_2}
$$

$$
\times \binom{x_3}{z_3} \left(\frac{a_3 \eta}{a_3 \eta + b_3} \right)^{z_3} \left(\frac{b_3}{a_3 \eta + b_3} \right)^{x_3 - z_3}
$$

$$
\times \binom{x_4}{z_4} \left(\frac{a_4 \eta}{a_4 \eta + b_4} \right)^{z_4} \left(\frac{b_4}{a_4 \eta + b_4} \right)^{x_4 - z_4} \mu^{z_1 + z_2 + \alpha_1 - 1} \eta^{z_3 + z_4 + \alpha_2 - 1}
$$

$$
\times \frac{\Gamma(z_1 + z_2 + z_3 + z_4 + x_5 + \alpha_1 + \alpha_2 + \alpha_3)}{\Gamma(z_1 + z_2 + \alpha_1) \Gamma(z_3 + z_4 + \alpha_2) \Gamma(x_5 + \alpha_3)} (1 - \mu - \eta)^{x_5 + \alpha_3 - 1}
$$

where the sum is taken over all possible sets of (z_1, z_2, z_3, z_4). In the case $(\mu, \eta) \in [\underline{\mu}, \overline{\mu}] \times [\underline{\eta}, \overline{\eta}]$, this kernel is easily bounded from below by

$$
\sum_{(z_1, z_2, z_3, z_4)} \binom{x_1}{z_1} \left(\frac{a_1 \underline{\mu}}{a_1 \overline{\mu} + b_1} \right)^{z_1} \left(\frac{b_1}{a_1 \overline{\mu} + b_1} \right)^{x_1 - z_1}
$$

$$
\times \binom{x_2}{z_2} \left(\frac{a_2 \underline{\mu}}{a_2 \overline{\mu} + b_2} \right)^{z_2} \left(\frac{b_2}{a_2 \overline{\mu} + b_2} \right)^{x_2 - z_2}
$$

$$
\times \binom{x_3}{z_3} \left(\frac{a_3 \underline{\eta}}{a_3 \overline{\eta} + b_3} \right)^{z_3} \left(\frac{b_3}{a_3 \overline{\eta} + b_3} \right)^{x_3 - z_3}
$$

$$
\times \binom{x_4}{z_4} \left(\frac{a_4 \underline{\eta}}{a_4 \overline{\eta} + b_4} \right)^{z_4} \left(\frac{b_4}{a_4 \overline{\eta} + b_4} \right)^{x_4 - z_4} \mu^{z_1 + z_2 + \alpha_1 - 1} \eta^{z_3 + z_4 + \alpha_2 - 1}
$$

$$
\times \frac{\Gamma(z_1 + z_2 + z_3 + z_4 + x_5 + \alpha_1 + \alpha_2 + \alpha_3)}{\Gamma(z_1 + z_2 + \alpha_1) \Gamma(z_3 + z_4 + \alpha_2) \Gamma(x_5 + \alpha_3)} (1 - \mu - \eta)^{x_5 + \alpha_3 - 1}
$$

and the parameters associated with the small set $[\underline{\mu}, \overline{\mu}] \times [\underline{\eta}, \overline{\eta}]$ are

$$
\epsilon = \left(\frac{a_1 \underline{\mu} + b_1}{a_1 \overline{\mu} + b_1} \right)^{x_1} \left(\frac{a_2 \underline{\mu} + b_2}{a_2 \overline{\mu} + b_2} \right)^{x_2} \left(\frac{a_3 \underline{\mu} + b_1}{a_3 \overline{\mu} + b_3} \right)^{x_3} \left(\frac{a_4 \underline{\mu} + b_4}{a_4 \overline{\mu} + b_4} \right)^{x_4}
$$

and

$$\nu(\mu,\eta) = \sum_{(z_1,z_2,z_3,z_4)} \binom{x_1}{z_1}\left(\frac{a_1\mu}{a_1\mu+b_1}\right)^{z_1}\left(\frac{b_1}{a_1\mu+b_1}\right)^{x_1-z_1}$$

$$\times \binom{x_2}{z_2}\left(\frac{a_2\mu}{a_2\mu+b_2}\right)^{z_2}\left(\frac{b_2}{a_2\mu+b_2}\right)^{x_2-z_2}$$

$$\times \binom{x_3}{z_3}\left(\frac{a_3\eta}{a_3\eta+b_3}\right)^{z_3}\left(\frac{b_3}{a_3\eta+b_3}\right)^{x_3-z_3}$$

$$\times \binom{x_4}{z_4}\left(\frac{a_4\eta}{a_4\eta+b_4}\right)^{z_4}\left(\frac{b_4}{a_4\eta+b_4}\right)^{x_4-z_4}$$

$$\times \frac{\Gamma(z_1+z_2+z_3+z_4+x_5+\alpha_1+\alpha_2+\alpha_3)}{\Gamma(z_1+z_2+\alpha_1)\Gamma(z_3+z_4+\alpha_2)\Gamma(x_5+\alpha_3)}$$

$$\times \mu^{z_1+z_2+\alpha_1-1} eta^{z_3+z_4+\alpha_2-1}(1-\mu-\eta)^{x_5+\alpha_3-1}$$

that is, the marginal (in (μ,η)) of the distribution

$$(z_1,z_2,z_3,z_4,\mu,\eta) \sim \mathcal{B}\left(x_1,\frac{a_1\mu}{a_1\mu+b_1}\right)\times\mathcal{B}\left(x_2,\frac{a_2\mu}{a_2\mu+b_2}\right)$$

$$\times\mathcal{B}\left(x_3,\frac{a_3\eta}{a_3\eta+b_3}\right)\mathcal{B}\left(x_4,\frac{a_4\eta}{a_4\overline{\eta}+b_4}\right)$$

$$\times\mathcal{D}(z_1+z_2+\alpha_1,z_3+z_4+\alpha_2,x_5+\alpha_3)\,.$$

TABLE 4.2. Small sets associated with the transition kernel $K((\mu,\eta),(\mu',\eta'))$ of the Multinomial Benchmark and corresponding renewal parameters. (The intervals on μ and ν are rescaled by a factor 10^2.)

State	1	2	3	4	5	6
$[\underline{\mu},\overline{\mu}]$	[.01, .8]	[.01, 2.4]	[.8, 2.3]	[2.4, 4.5]	[4.8, 7.1]	[.01, 2.9]
$[\underline{\eta},\overline{\eta}]$	[.01, 1.3]	[1.6, 3.2]	[.01, 1.4]	[.01, 2.6]	[.01, 2.4]	[3.2, 5.9]
ϵ	.799	.642	.728	.603	.603	.544
ϱ	.0093	.0081	.010	.010	.0103	.0098

A preliminary run of the Gibbs sampler leads to select the sets given in Table 4.2 as those maximizing the probability of renewal. Table 4.2 shows that these probabilities are much smaller than for the two first examples, another occurrence of the "curse of dimensionality". However, the number of renewals between two meetings of two given chains is only 5.2. Figure 4.7 gives a summary of the behaviour of the Gibbs sampler in this setting, with a example of the paths of two independent chains until they meet.

Using the reduction device of §4.4.5, with two chains starting from states "1" and "4", Figure 4.8(a) shows that 50,000 encounters for the discretized

FIGURE 4.7. Description of the posterior distribution of (μ, η) from the sample obtained by Gibbs sampling in the Multinomial benchmark. The figure on the upper right side provides the steps of both Markov chains used to implement the divergence criterion between two encounters.

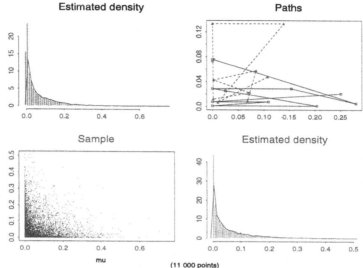

Markov chains are not sufficient to ensure convergence of the six selected divergences, while Figure 4.8(b) exhibits a sufficient stability to conclude about convergence after $500,000$ encounters. Note that, given that the probability of renewal is 0.0475, and there are an average of 5.498 renewals between two encounters, the production of $500,000$ encounters for the discretized Markov chains requires $57,878,927$ iterations of the Gibbs sampler, which is rather a stupendous number on a chain that simple!

FIGURE 4.8. Convergence of six selected divergences based on two discretized chains $(\mu^{(t)}, \eta^{(t)})$ started from states $[.0001, .008] \times [.0001, .013]$ ("1") and $[.024, .045] \times [.0001, .024]$ ("4") in the Multinomial Benchmark. (a) The number of encounters of the discretized chains is $50,000$. (b) The number of encounters of the discretized chains is $500,000$.

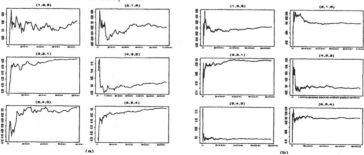

4.5.4   Comments

Potential difficulties with our method are that

1. it requires the determination of small sets A_j with manageable associated parameters (ϵ_j, ν_j);

2. it provides conservative convergence assessments for the discretized Markov chain.

The latter point is a consequence of our rigorous requirements for the Markov structure of the discrete chain and exact convergence of the divergence approximations. It is thus rather comforting to exhibit safe bounds on the number of simulations necessary to give a good evaluation of the distribution of interest. We indeed prefer our elaborate and slow but theoretically well-grounded convergence criterion to handy and quick alternatives with limited justifications because the latter are tested in very special setups but usually encounter difficulties and inaccuracies outside these setups. Note, however, that the convergence assessment does not totally extend to the original continuous Markov chain. Moreover, the few examples treated here show that the use of the estimate of IP as a control variate technique leads to long delays in the convergence diagnostic.

The first difficulty is obviously more of a concern but there are theoretical assurances that small sets exist in most MCMC setups. As shown in §4.2.2, Mykland *et al.* (1995) have proposed some quasi-automated schemes to construct such small sets by hybrid modifications of the original MCMC algorithm. Note also that the techniques we used in the examples of §4.5, namely to bound from below the conditional distributions depending on θ, can be reproduced in many Gibbs setups and in particular for *Data Augmentation*. At last, the choice of the space where the small sets are constructed is open and, at least for Gibbs samplers, there are often obvious choices as in the Duality Principle of §1.5. (See also Guihenneuc-Jouyaux and Robert, 1998, for an illustration on an $AR(1)$ changepoint model.) It must be pointed out, however, that missing data structures like the mixtures of distributions (see §3.4) are notorious[2] for leading to very small bounds in the minorization condition (4.1), and that other convergence diagnostics based on the natural finite Markov chains generated in these setups would be preferable. Moreover, as also pointed out by Gilks, Roberts and Sahu (1998) for an acceleration method using regeneration, the applicability of the method is limited in high-dimensional problems by the difficulty to obtain efficient minorizing conditions[3], even though new developments are bound to occur in this area, given the current interest.

---

[2] It actually happens that all the applications in the following chapters belong to this class of models and thus prevent the application of the divergence criterion.

[3] As pointed out by Herman van Dijk (1997, personal communication), the shape of the small sets in multidimensional settings is bound to influence the

4.6 Renewal theory for variance estimation

4.6.1 Estimation of the asymptotic variance

Renewal theory can also provide an implementable estimation procedure for the asymptotic variance σ_h^2 and thus a convergence criterion for MCMC algorithms. Indeed, since the random variables $(z_t - \mu_t \mathbb{E}^{\tilde{\pi}}[h(x)])$ in (4.6) are independent, the renewal variance $\tilde{\sigma}_A^2$ can be estimated by the usual sum of squares estimator

$$\frac{1}{T} \sum_{t=1}^{T} (S_t - \lambda_t \mathbb{E}^{\tilde{\pi}}[h(x)])^2$$

or, since the expectation $\mathbb{E}^{\tilde{\pi}}[h(x)]$ is unknown, by

$$\hat{\sigma}_A^2 = \frac{1}{T} \sum_{t=1}^{T} \left(S_t - \lambda_t \sum_{\ell=1}^{T} S_\ell / N \right)^2. \tag{4.16}$$

We then deduce the following invariance property.

Proposition 4.6.1 *For every small set A such that (4.1) holds, the ratio*

$$\frac{\hat{\sigma}_A^2 T}{N} \tag{4.17}$$

converges a.s. (in N) to σ_h^2.

Proof. The result follows immediately from the a.s. convergence of $\hat{\sigma}_A^2$ to $\tilde{\sigma}_A^2$ and of (ii) in Lemma 4.2.1, as N/T converges a.s. to μ_A. Since

$$\frac{1}{\sqrt{N}} \sum_{n=1}^{N} \left(h(x_n) - \mathbb{E}^{\tilde{\pi}}[h(x)] \right) - \sqrt{\frac{T}{N}} \frac{1}{\sqrt{T}} \sum_{t=1}^{T} (S_t - \lambda_t \mathbb{E}^{\tilde{\pi}}[h(x)])$$

converges a.s. to 0 and the second term converges in distribution to $\mathcal{N}(0, \tilde{\sigma}_A^2/\mu_A)$, while the first term converges to $\mathcal{N}(0, \sigma_h^2)$ if $\sigma_h > 0$, the asymptotic invariance of the ratio $\hat{\sigma}_A T/N$ follows.   □□

That (4.17) is a convergent estimator of σ_h^2 is obviously an interesting feature, since it shows that renewal theory can lead to the estimation of the asymptotic variance, just as well as spectral theory or other time-series methods. However, the main incentive for using renewal theory is for us the asymptotic invariance of this ratio, as we can deduce a convergence

---

performance of the method. In this regard, the usual choice of hypercubes is most likely subefficient.

criterion: *given several small sets* A_i, *wait until the ratios* $\sigma^2_{A_i} T_i/N$ *have similar values.*

Although the theoretical basis of this method is quite sound, we are faced with two implementation caveats: first, the criterion is conservative, in the sense that it requires the slowest ratio to converge for the algorithm to stop. Second, as in other parallel methods, the dependence on the starting values is crucial, since close A_i's will lead to earlier terminations than far spaced A_i's, while it is usually impossible to assess how close "close" is. However, we will introduce below a general class of models for which these drawbacks can be overcome.

4.6.2   Illustration for the Cauchy Benchmark

As shown in §4.5.2, the set $A = [x_2 - r, x_2 + r]$ is a small set for the Markov chain on the θ's. In addition,

$$\epsilon = \frac{1 + \rho^2_{11}}{1 + \rho^2_{12}} \frac{1}{1 + \rho^2_{22}} \frac{1 + \rho^2_{31}}{1 + \rho^2_{32}}$$

can be easily computed. Table 4.3 shows that ϵ decreases quite slowly to 0 as r increases. The number of renewals in a sequence of Gibbs iterations is thus likely to be sufficiently high although both quantities are not strongly connected. The average number of steps between two returns to A goes as low as 8.8 when $r = .54$. Note the actual stabilization to $\sigma^2_h \simeq 1150$ for most values of r. This large variance factor may be explained by the Cauchy tails of the posterior distribution as well as the iterative switching from one mode to another in the Gibbs algorithm.

TABLE 4.3.   Renewal parameters when $A = [x_2 - r, x_2 + r]$ and $h(x) = x$ for $x_1 = -8$, $x_2 = 8$, $x_3 = 17$ and $\sigma^2 = 100$ in the Cauchy Benchmark (based on 1,000,000 simulations); $\bar{\tau}_A$ is the mean excursion time and $\hat{\sigma}^2_A$ is the estimate of σ^2_h based on (4.17). (*Source:* Robert, 1995.)

r	.1	.21	.32	.43	.54	.65	.76	.87	.98	1.09
ϵ_A	.92	.83	.73	.63	.53	.45	.38	.31	.26	.22
$\bar{\tau}_A$	25.3	13.9	10.5	9.6	8.8	9.6	9.8	10.4	11.4	12.7
$\hat{\sigma}^2_A$	1135	1138	1162	1159	1162	1195	1199	1149	1109	1109

4.6.3   Finite Markov chains

Finite state space Markov chains are, again, ideal settings for the application of the above technique. Consider a chain $(x^{(t)})$ with values in the finite space $\mathcal{X} = \{1, \ldots, m\}$ and transition matrix $\mathbb{P} = (p_{ij})$. The cardinal m is often of the form k^p, in particular in missing data problems. We assume the chain $(x^{(t)})$ to be irreducible and aperiodic. Define $\pi = (\pi_1, \ldots, \pi_m)$ as the stationary distribution and take π_{i_0} as the probability of the most probable state, i_0. Then, if $A = \{i_0\}$, renewal theory applies with $\epsilon = 1$ and $\nu = (p_{i_0 1}, \ldots, p_{i_0 m})$. The 'perturbation' (4.4) is then useless and the ratio $K(x^{(t)}, x^{(t+1)})/\nu(x^{(t+1)})$ is equal to 1.

The consequences of this simplification are, however, far from trivial. Indeed, the sums $\sum_{w=1}^{W} h(x^{(w)})$ can then be decomposed into iid sums

$$S_j = \sum_{w=\tau_j+1}^{\tau_{j+1}} h(x_w) = h(t_{i_0}) + \sum_{w=\tau_j+1}^{\tau_{j+1}-1} h(x_w) \qquad (j = 1, \ldots),$$

where

$$\tau_j = \inf\{\omega > \tau_{j-1}; x^{(\omega)} = i_0\}.$$

The variance σ_h^2 of the asymptotically normal expression

$$W^{-1/2} \sum_{w=1}^{W} (h(x_w) - \mathbb{E}^{\pi}[h(x)])$$

can therefore be estimated directly by (4.16) and (4.17). Moreover, a convergence criterion can be derived by considering other probable states $i_1, \ldots, i_c$ and checking convergence to the same value of the corresponding estimators (4.17) of σ_h^2.

Consider for instance the special case $\mathcal{X} = \{0, 1, 2, 3\}$ and $(x^{(t)})$ with transition matrix

$$\mathbb{P} = \begin{pmatrix} .26 & .04 & .08 & .62 \\ .05 & .24 & .03 & .68 \\ .11 & .10 & .08 & .71 \\ .08 & .04 & .09 & .79 \end{pmatrix}.$$

The stationary distribution is $\pi = (.097, .056, .085, .762)$ and the corresponding mean is 2.507. If we use a simulation of this Markov chain, the four states can be chosen as renewal sets and an estimate of σ_h^2 can be constructed for each state, based on (4.16) and (4.17). Table 4.4 provides the different estimates of σ_h^2, for 5000 and 500,000 simulations from P, for $h(x) = x$. The larger simulation study clearly shows that convergence is achieved since the four estimates of σ_h^2 are equal up to the second decimal.

In most practical setups, the most probable state i_0 is unknown beforehand, as are the other probable states $i_1, \ldots, i_c$. We suggest to take advantage of some 'burn-in' initial iterations of the MCMC algorithm to derive these states or some approximations since, when the chain is close to stationarity, the most commonly sampled states are the most probable states for π. In some setups such as mixtures or Ising models, the state

TABLE 4.4. Estimates of σ_h^2 for $h(x) = x$, based on renewal at states i_0. (*Source:* Robert, 1995.)

$n\backslash i_0$	0	1	2	3
5000	1.19	1.29	1.26	1.21
500,000	1.344	1.335	1.340	1.343

space is too large for a single state to be probable enough, i.e. to have a probability of occurrence larger than .01 or .005 say. In this case, the renewal set A can be selected as an union of states, $A = \{i_0, i_1, \ldots, i_r\}$, the renewal measure ν being defined by

$$\nu(i) \propto \min_{j \in A} p_{ji} \qquad (4.18)$$

and the bound ϵ by

$$\epsilon = \sum_{i=1}^{r} \min_{j \in A} p_{ji}. \qquad (4.19)$$

If r is too large for the above distribution to be computed exactly, note that (4.19) is bounded from below by a similar sum on the most common states and that an additional Metropolis step can be used to simulate from (4.18).

Example 4.6.1 The finite Markov chain $(z_1^{(t)}, z_2^{(t)}, z_3^{(t)}, z_4^{(t)})$ has a support
MULTINOMIAL of cardinal $(x_1 + 1) \times \ldots \times (x_4 + 1) = 7280$ but some configurations have
BENCHMARK a much higher frequency than the others. Among the most frequent ones, consider $(0, 0, 0, 0)$, $(0, 1, 1, 0)$ and $(1, 1, 1, 1)$ with respective frequencies of occurrence .045, .019 and .003. The estimates of σ_h^2 for $h(\mu, \eta) = \eta\mu$ are given in Table 4.5 for different numbers of iterations. The agreement between the three estimates thus requires as much as 1 million iterations, but the small renewal probability of state $(1, 1, 1, 1)$ must be accounted for, since the estimates for states $(0, 0, 0, 0)$ and $(0, 1, 1, 0)$ already agree for 500,000 iterations.

Table 4.6 reproduces the experiment for the Rao-Blackwellized versions of the estimates, i.e. for

$$\begin{aligned}
S_t &= \sum_{n=\tau_t+1}^{\tau_{t+1}} \mathbb{E}[h(\eta, \mu)|z_1^{(n)}, z_2^{(n)}, z_3^{(n)}, z_4^{(n)}] \\
&= \frac{(z_1^{(n)} + z_2^{(n)} + 0.5)(z_3^{(n)} + z_4^{(n)} + 0.5)}{(x_5 + z_1^{(n)} + z_2^{(n)} + z_3^{(n)} + z_4^{(n)} + 2.5)} \\
&\quad \times (x_5 + z_1^{(n)} + z_2^{(n)} + z_3^{(n)} + z_4^{(n)} + 1.5)^{-1}
\end{aligned}$$

TABLE 4.5. Estimates of σ_h^2 for $h(\mu, \eta) = \mu\eta$, based on renewal at state j for the missing data Markov chain $(z^{(t)})$.

$n\backslash j$	$(0,0,0,0)$	$(0,1,1,0)$	$(1,1,1,1)$
5000	.00427	.00452	.00466
10,000	.00398	.00423	.00467
50,000	.00453	.00429	.00420
100,000	.00434	.00433	.00462
500,000	.00432	.00426	.00409
1,000,000	.00429	.00433	.00426

It shows that the agreement somehow occurs faster, i.e. between $500,000$ and 1 million iterations for the three states, and that it is much faster for the two main states, since $10,000$ iterations show already a strong closeness between both estimates.   ‖

TABLE 4.6. Estimates of σ_h^2 for the sums of the conditional expectations $\mathbb{E}[\mu\eta | z_1^{(t)}, z_2^{(t)}, z_3^{(t)}, z_4^{(t)}]$, based on renewal at state j for the missing data Markov chain $(z^{(t)})$.

$n\backslash j$	$(0,0,0,0)$	$(0,1,1,0)$	$(1,1,1,1)$
5000	.00150	.00157	.00134
10,000	.00158	.00159	.00168
50,000	.00190	.00177	.00142
100,000	.00190	.00189	.00167
500,000	.00187	.00188	.00174
1,000,000	.00189	.00191	.00188

5
Control by the Central Limit Theorem

Didier Chauveau
Jean Diebolt
Christian P. Robert

5.1 Introduction

Distinctions between single chain and parallel chain control methods have already been discussed in Chapter 2. However, as Brooks and Roberts (1998) point out, other characteristics must be taken into account for evaluating control methods. An important criterion is the *programming investment*: diagnostics requiring problem-specific computer codes for their implementation (e.g., requiring knowledge of the transition kernel of the Markov chain) are far less usable for the end user than diagnostics solely based upon the outputs from the sampler, which can use available *generic codes*. Another criterion is *interpretability*, in the sense that a diagnostic should preferably require no interpretation or experience from the user.

This chapter is mainly devoted to a methodological approach which makes use of parallel chains, and is based on the following *shift of paradigm*: instead of checking for stationarity of $(x^{(t)})$, we primarily aim at controlling the precision of estimates like (1.2). A natural way to do this is through confidence regions based on normal approximation resulting from the Central Limit Theorem (CLT) for Markov chains. This approach results in control techniques which comply with the above criterion, i.e. they are not problem-specific and they provide automated diagnostics.[1]

Section 5.2 presents the CLT for MCMC algorithms, and connections between the CLT and renewal theory in discrete situations, already introduced in §4.2. Section 5.3 describes two control methods based on the CLT through the above shift of paradigm, while Section 5.4 extends this methodology to continuous state space Markov chains. Section 5.5 applies these methods to some benchmark examples, and Section 5.6 suggests an alternative use of the CLT for control purpose mainly in latent variable models, based on Robert, Rydén and Titterington (1998).

---

[1] These control tools have been included in a generic computer code which is available at http://www-math.univ-mlv.fr/users/chauveau/cltc.html

5.2   CLT and Renewal Theory

5.2.1   Renewal times

We consider here a finite irreducible aperiodic Markov chain $(x^{(t)})$ with state space E. For each subset A of E, we denote by

$$N_n(A) = \sum_{t=1}^{n} \mathbb{I}\left(x^{(t)} \in A\right) \qquad (5.1)$$

the occupation time of A during the first n steps. For each real function h defined on the state space, consider

$$S_n(h) = \sum_{t=1}^{n} h\left(x^{(t)}\right) \quad \text{and} \quad S_n(\bar{h}) = \sum_{t=1}^{n} \left[h\left(x^{(t)}\right) - \mathbb{E}^f[h]\right],$$

where f denotes, as in previous chapters, the density of the invariant distribution of the Markov chain. We will make use of *Wald's equation*:

Theorem 5.2.1 *Let* $Z_1, Z_2, \ldots$ *be iid random variables such that* $\mathbb{E}[Z_1^2] < \infty$, *and* T *be a stopping time for* $(Z_t)_{t \geq 1}$ *such that* $\mathbb{E}[T] < \infty$. *Then*

(i) $\mathbb{E}\left[\sum_{p=1}^{T} Z_p\right] = \mathbb{E}[T]\,\mathbb{E}[Z_1]$

(ii) $\text{var}\left[\sum_{p=1}^{T} Z_p\right] = \mathbb{E}[T]\,\text{var}[Z_1]$

Wald's theorem for square integrable martingales can be found in Dacunha-Castelle and Duflo (1986, p. 96).

We assume for simplicity that the Markov chain starts from $x^{(0)} = i$. When $x^{(0)}$ is generated from a general initial distribution μ_0, we only have to shift the starting time to the first time $x^{(t)} = i$. Let

$$T_i(1) = \inf\{t > 0 : x^{(t)} = i\}$$

be the first time $t > 0$ the chain returns to the state i, and $T_i(0) = 0$ by convention. The r.v. $T_i(1)$ is a stopping time with respect to the sequence $(x^{(t)})_{t \geq 0}$. Define the stopping time $T_i(p)$, $p \geq 2$, as the pth return time to state i. Let $\tau_i(p)$, $p \geq 1$, be the duration of the pth excursion out of state i. The $\tau_i(p)$'s and $T_i(p)$'s are connected by $T_i(1) = \tau_i(1)$ and

$$T_i(p) = T_i(p-1) + \tau_i(p), \quad p \geq 1.$$

Proposition 5.2.2 *For any* $i \in E$, *the* $\tau_i(p)$ *'s,* $p \geq 1$, *are iid and have finite moments of all orders.*

Proposition 5.2.2 can be found, e.g., in Chung (1967). Note that it holds for any starting distribution by considering only the $\tau_i(p)$'s for $p \geq 2$. Let $q_i(t)$ be the random number of returns to state i before time t,

$$q_i(t) = \max\{p \geq 1 : T_i(p) \leq t\}.$$

We have

$$q_i(t) + 1 = \sum_{s=0}^{t} \mathbb{I}(x^{(s)} = i);$$

it follows that

$$\mathbb{E}_i[q_i(t) + 1] = \sum_{s=0}^{t} p_{i,i}^{(s)};$$

therefore,

$$\lim_{t \to \infty} \mathbb{E}_i \left[\frac{q_i(t) + 1}{t + 1} \right] = \pi_i. \qquad (5.2)$$

We define the block sums

$$Z_p(h) = \sum_{t=T_i(p)+1}^{T_i(p+1)} h(x^{(t)}) \quad \text{and} \quad Z_p(\bar{h}) = \sum_{t=T_i(p)+1}^{T_i(p+1)} \left[h(x^{(t)}) - \mathbb{E}^f[h] \right],$$

for $p \geq 0$, corresponding to sums over the excursions out of i.

Proposition 5.2.3 *Let the finite state Markov chain $(x^{(t)})_{t \geq 0}$ starts from $x^{(0)} = i$. Then for any h the $Z_p(h)$'s, $p \geq 0$, are iid random variables and have finite moments of all orders.*

Proposition 5.2.3 can be found in, e.g., Chung (1967). It also holds for any starting distribution by considering the $Z_p(h)$'s for $p \geq 1$. The following theorem states a Strong Law of Large Numbers for finite irreducible and aperiodic Markov chains (see, e.g., Dacunha-Castelle and Duflo, 1986):

Theorem 5.2.4 *If the finite Markov chain is irreducible and aperiodic, then for any initial distribution μ_0,*

(i) $\mathbb{E}_i[\tau_i(1)] = \mathbb{E}_{\mu_0}[\tau_i(p)] = \pi_i^{-1}$ *for* $p \geq 2$.

(ii) $\lim_{n \to \infty} \dfrac{S_n(h)}{n} = \mathbb{E}^f[h]$ *a.s.*

As a consequence,

$$\lim_{n \to \infty} \frac{N_n(i)}{n} = \lim_{n \to \infty} \frac{q_i(n) + 1}{n + 1} = \pi_i \quad \text{a.s.} \qquad (5.3)$$

5.2.2   CLT for finite Markov chains

Recall that $S_n(\bar{h}) = \sum_{t=1}^{n}[h(x^{(t)}) - \mathbb{E}^J[h]]$. The variance of the random variables $n^{-1/2} S_n(\bar{h})$ converges to a limiting variance

$$\sigma^2(h) = \lim_{n \to \infty} n^{-1} \operatorname{var}_{\mu_0}[S_n(\bar{h})].$$ (5.4)

(See, e.g., Kemeny and Snell, 1960.)

Theorem 5.2.5 *If the finite Markov chain is irreducible and aperiodic, then for any initial distribution μ_0,*

(i) $\operatorname{var}_i[Z_0(h)] = \operatorname{var}_{\mu_0}[Z_p(h)] = \dfrac{\sigma^2(h)}{\pi_i}$ *for* $p \geq 1.$

(ii) $\dfrac{S_n(\bar{h})}{\sigma(h)\sqrt{n}} \overset{\mathcal{L}}{\rightsquigarrow} \mathcal{N}(0,1)$ *as* $n \to \infty.$

Proof. It suffices to prove the result for nonnegative h's and to assume that $x^{(0)} = i$. Since $T_i(q_i(t)) \leq t < T_i(q_i(t) + 1)$, it follows that

$$0 \leq T_i(q_i(t) + 1) - t < \tau_i(q_i(t) + 1).$$

Since $h \geq 0$,

$$\sum_{p=0}^{q_i(t)} Z_p(h) \leq S_t(h) \leq \sum_{p=0}^{q_i(t)+1} Z_p(h).$$

Therefore,

$$\left| S_t(h) - \sum_{p=0}^{q_i(t)} Z_p(h) \right| \leq \|h\|_\infty \tau_i(q_i(t) + 1).$$

Hence,

$$\left| S_t(\bar{h}) - \sum_{p=0}^{q_i(t)} Z_p(\bar{h}) \right| \leq C \tau_i(q_i(t) + 1),$$ (5.5)

where C is an appropriate constant. It follows from (5.4) and (5.5) that

$$\lim_{t \to \infty} t^{-1} \operatorname{var}_i \left[\sum_{p=0}^{q_i(t)} Z_p(\bar{h}) \right]$$

$$= \lim_{t \to \infty} t^{-1} \operatorname{var}_i[S_{T_i(q_i(t)+1)}(\bar{h})] = \sigma^2(h).$$ (5.6)

Let σ_Z^2 denote the common variance of the $Z_p(\bar{h})$'s. The event $\{q_i(t) + 1 = n\}$ is $(\tau_i(1), \dots, \tau_i(n))$-measurable or, equivalently, $(Z_0(h), \dots, Z_{n-1}(h))$-measurable. We apply Wald's equation (Theorem 5.2.1) for the iid random variables $Z_p(h)$:

$$\operatorname{var}_i[Z_0(\bar{h}) + \cdots + Z_{q_i(t)}(\bar{h})] = \operatorname{var}_i[S_{T_i(q_i(t)+1)}(\bar{h})]$$

$$= \operatorname{var}_i[Z_0(\bar{h})] \, \mathbb{E}_i[q_i(t) + 1].$$ (5.7)

In view of (5.3), (5.6) and (5.7), we have

$$\sigma^2(h) = \lim_{t \to \infty} \text{var}_i \left[\frac{S_{T_i(q_i(t)+1)}(\bar{h})}{t} \right] = \sigma_Z^2 \, \pi_i,$$

implying *(i)*. The proof of *(ii)* relies on a CLT for a random number of summands (see, e.g., Billingsley, 1986, p. 380), applied to the $Z_p(\bar{h})$'s for $1 \leq p \leq q_i(n) - 1$. It makes use of (5.5).   □□

5.2.3   More general CLTs

For general state spaces, we will state a CLT which can be applied, for instance, when the Markov chain is geometrically ergodic. The basic ideas of this extension are, first, to extend the previous results (Theorems 5.2.4 and 5.2.5) to atomic Markov chains (the renewal state i being replaced with an atom A) and second, to transform general Markov chains to atomic Markov chains by splitting a small set (see §4.2). This extension requires square integrable h's, return times $\tau_A(p)$, $p \geq 1$, and block sums

$$Z_p(\bar{h}) = \sum_{t=T_A(p)+1}^{T_A(p+1)} \left[h \left(x^{(t)} \right) - \mathbb{E}^f[h] \right], \quad p \geq 0.$$

The following theorem can be found, e.g., in Robert (1996, p. 123):

Theorem 5.2.6 *Let $(x^{(t)})_{t \geq 0}$ be a Harris recurrent Markov chain with invariant probability density f. Assume that there exist a finite function V, a function $g \geq 1$ such that $\mathbb{E}^f[g^2] < \infty$, a small set C and a constant $0 < b < \infty$ such that*

$$\int_E V(y)P(x, dy) - V(x) \leq -g(x) + b\mathbb{I}_C(x) \quad \text{for all} \quad x \in E.$$

Then for all h's such that $|h| \leq g$, the variances of the random variables $S_n(\bar{h})/\sqrt{n}$ converge to a finite limiting variance

$$\sigma^2(h) = var^f \left[h(x^{(0)}) \right] + 2 \sum_{s=1}^{\infty} cov^f \left[h(x^{(0)}), h(x^{(s)}) \right] \geq 0.$$

If in addition $\sigma^2(h) > 0$ then

$$\frac{S_n(\bar{h})}{\sigma(h)\sqrt{n}} \overset{\mathcal{L}}{\rightsquigarrow} \mathcal{N}(0, 1) \quad \text{as} \quad n \to \infty.$$

If $\sigma^2(h) = 0$ then $S_n(\bar{h})/\sqrt{n}$ converges a.s. to 0.

5.3   Two control methods with parallel chains

Since we want to use a normal approximation, our main goal is to estimate the time needed to reach approximate normality for suitable functions of $(x^{(t)})$. We propose here to use statistical normality tests on the normalized sums

$$\frac{1}{\sqrt{n}} \sum_{t=1}^{n} \left(h(x^{(t)}) - \mathbb{E}^J[h] \right),$$
(5.8)

with samples obtained from parallel chains, and to monitor variance stabilization near the limiting variance appearing in the CLT.

We first investigate the case of finite state Markov chains. The motivations for adopting this point of view have already been presented in §3.1, and valid techniques for linking finite and continuous chains (see the Duality Principle in §1.5, and the discretization method of Chapter 4) have been discussed. In our setup, we will see in addition that for finite chains, the limiting variance in the CLT can be consistently estimated and compared with another estimate of the variance after n iterations, giving a helpful control variate which may be partially extended to the continuous case.

5.3.1   CLT and Berry-Esséen bounds for finite chains

Consider a regular Markov chain $(x^{(t)})$ with finite state space

$$E = \{1, \ldots, I\},$$

transition matrix $\mathbb{P}$ and stationary distribution $\pi = (\pi_1, \ldots, \pi_I)$. Our goal is to obtain reliable estimates and confidence intervals for the stationary probabilities π_j, $1 \leq j \leq I$, using the normal approximation. The main tool used for this purpose is a CLT on the time spent in a given state during the first n steps of an ergodic Markov chain, with the limiting variance available in closed form using $\mathbb{P}$ and π, as given, e.g., in Kemeny and Snell (1960). This setup has already been introduced in §4.4.1, where two matrices of interest have been defined: the matrix A with all rows equal to π, and the fundamental matrix

$$\mathbb{Z} = \left(I - (\mathbb{P} - A) \right)^{-1} = I + \sum_{k=1}^{\infty} (\mathbb{P}^k - A).$$
(5.9)

The limiting variance in the CLT depends on $\mathbb{Z}$ in the following sense: let h and g be two real-valued functions defined on E (considered as column vectors). The limiting covariance matrix is the $I \times I$ symmetric matrix $C = (c_{ij})$ such that, for any starting distribution $\tilde{\pi}$,

$$\lim_{n \to \infty} \frac{1}{n} \text{cov}_{\tilde{\pi}} \left[\sum_{t=1}^{n} h(x^{(t)}), \sum_{t=1}^{n} g(x^{(t)}) \right] = h^T C g$$
(5.10)

$$= \sum_{i,j=1}^{I} h(i) c_{ij} g(j) \tag{5.10}$$

Note that (5.10) is stated in Kemeny and Snell (1960) with π as the starting distribution to keep computations simple. However, (5.10) holds for any starting distribution $\tilde{\pi}$. The matrix C is related to $\mathbb{Z} = (z_{ij})$ and π by

$$c_{ij} = \pi_i z_{ij} + \pi_j z_{ji} - \pi_i \delta_{ij} - \pi_i \pi_j, \tag{5.11}$$

where $\delta_{ij} = 0$ for $i \neq j$ and $\delta_{ii} = 1$. For each state $i \in E$, let

$$N_n(i) = \sum_{t=1}^{n} \mathbb{I}_i(x^{(t)})$$

denote, as in §5.2.1, the occupation time of state i during the first n steps. Specializing (5.10) to the indicator functions $h = \mathbb{I}_i$ and $g = \mathbb{I}_j$ leads to

$$\lim_{n \to \infty} \frac{1}{n} \mathrm{cov}_{\tilde{\pi}} [N_n(i), N_n(j)] = c_{ij}$$

and

$$\lim_{n \to \infty} \frac{1}{n} \mathrm{var}_{\tilde{\pi}} [N_n(i)] = c_{ii}.$$

For any function $h : E \to \mathbb{R}$, consider the quantities

$$\sigma_n^2(h) = \frac{1}{n} \mathrm{var}_{\tilde{\pi}} [S_n(h)], \quad \sigma^2(h) = \lim_{n \to \infty} \sigma_n^2(h) = h^T C h. \tag{5.12}$$

Here $\sigma_n^2(h)$ denotes the variance after n steps starting from $x^{(0)} \sim \tilde{\pi}$, and $\sigma^2(h)$ denotes the corresponding limiting variance. The Central Limit Theorem for Markov chains introduced in §4.2.3 and §5.2.2 (Theorem 5.2.5), when applied to $h = \mathbb{I}_i$, leads to a CLT for the occupation times:

$$\left(\frac{N_n(1) - n\pi_1}{\sqrt{n}}, \ldots, \frac{N_n(I) - n\pi_I}{\sqrt{n}} \right) \overset{\mathcal{L}}{\rightsquigarrow} \mathcal{N}(0, C).$$

The problem of the time required for the normal approximation to be valid addresses the question of the convergence rate in the CLT. In good settings, upper bounds for this rate are given by the Berry-Esséen Theorem for Markov chains. The Berry-Esséen Theorem is said to hold when

$$\sup_{x \in \mathbb{R}} \left| \mathbb{P}_{\tilde{\pi}} \left[\frac{S_n(\bar{h})}{\sigma(h) \sqrt{n}} \leq x \right] - \Phi(x) \right| = \mathcal{O}(n^{-1/2}). \tag{5.13}$$

General conditions have been given for (5.13) to hold in both the discrete and continuous cases (see, e.g., Bolthausen 1982). However, a workable bound requires precise estimation of the constant involved in the right-hand side of (5.13). This question has been investigated by Mann (1996) and Lezaud (1998) for countable state chains, but the proposed constants are far too large for practical use in our case. Moreover, computing these bounds requires knowledge of unavailable quantities (e.g., the *gap* of the transition kernel).

Another approach consists in using the Berry-Esséen Theorem for the iid case (Feller, 1971), where the constant has been precisely evaluated. If $X_i, \ldots, X_n$ are iid random variables with zero expectation, variance σ^2, and such that $\rho = \mathbb{E}\big[|X|^3\big] < \infty$, then

$$\sup_{x \in \mathbb{R}} \left| \mathbb{P}\left[\frac{\sum_{t=1}^{n} X_i}{\sigma \sqrt{n}} \leq x \right] - \Phi(x) \right| < C_{BE} \frac{\rho}{\sigma^3 \sqrt{n}},$$

where the constant, initially evaluated at $33/4$, has been lowered down to $C_{BE} \leq 0.7915$ (see, e.g., Seoh and Hallin, 1997). This approach can be transposed to the case of Markov chains with the help of renewal theory, through the iid random variables $Z_p(h)$ defined in §5.2.1. Typically, if $h = \mathbb{I}_A$ for a subset $A \in E$, then

$$\sup_{x \in \mathbb{R}} \left| \mathbb{P}_{\tilde{\pi}}\left[\frac{\sqrt{\pi_i} S_{T_i(q)}(\bar{h})}{\sigma(h)\sqrt{q}} \leq x \right] - \Phi(x) \right| \leq \frac{C_{BE} \pi_i^{3/2} \mathbb{E}\big[\tau_i^3\big]}{\sigma^3(h)\sqrt{q}} . \tag{5.14}$$

However, Chauveau and Diebolt (1997) show that this approach is helpless in practical situations. This is mainly due to the poor quality of the Berry-Esséen bound.

5.3.2   *Convergence assessment by normality monitoring*

The proposed convergence assessments are basically derived from the discrete setting given in §5.3.1. They are based on the asymptotic behaviour of $\sigma_n^2(h)$ and $S_n(\bar{h})/\sqrt{n}$ for $h = \mathbb{I}_i$, $i \in E$, or more generally $h = \mathbb{I}_A$ for a collection of subsets $A \subset E$. If stationarity is reached for $(x^{(t)})$, then the variance after n steps, $\sigma_n^2(h)$, should be close to the limiting variance $\sigma^2(h)$ and the distribution of $S_n(h)/\sqrt{n}$ should be approximately normal. Since we are more interested in normality than in stationarity assessment, we propose two complementary methods to guarantee that the CLT can effectively be used after n steps of the algorithm under consideration, to build reliable confidence intervals for a class of normalized sums $S_n(h)/\sqrt{n}$. The first method is based on normality assessment and the second one monitors variance stabilization. Both methods use independent parallel chains started from a suitably dispersed distribution $\tilde{\pi}$. (See the debate in Gelman and Rubin, 1992, and Geyer, 1992, about the feasibility of this requirement.)

Basically, the normality control method consists in running m parallel chains started from some preassigned distribution, and testing a normality hypothesis H_0 for the r.v.'s $N_n(i)$, $i \in I$, at arbitrary selected times $n_1 < n_2 < \cdots < n_k < \cdots$, until acceptance of normality. Consider first a single state $i \in I$. Define

$$N_n^{(\ell)}(i) = \sum_{t=1}^{n} \mathbb{I}\left(x_\ell^{(t)} = i\right), \quad 1 \leq \ell \leq m,$$

the occupation time of state i for chain ℓ during the first n steps. We propose to check approximate normality using the Shapiro-Wilk test (Shapiro and Wilk, 1965) with a preassigned significance level α to be tuned. This test consists in a comparison between the maximum likelihood estimator of the variance, and the variance best linear unbiased estimator under the null hypothesis. The Shapiro-Wilk test statistic SW belongs to $(0,1)$, and assumes values close to 1 if the null hypothesis H_0 is true (see, e.g., Capéraà and van Cutsem, 1988). This test does not require prior knowledge of the expectation $n\pi_i$ of $N_n(i)$. For m chains with initial distribution $\tilde{\pi}$, and arbitrary increasing times $n_0 = 0 < n_1 < n_2 < \cdots$, the control method starts with $k = 1$ and proceeds as follows:

1. **Run the m chains for $(n_k - n_{k-1})$ more iterations.**

2. **Update the sample** $\left(\dfrac{N_{n_k}^{(1)}(i)}{\sqrt{n_k}}, \ldots, \dfrac{N_{n_k}^{(m)}(i)}{\sqrt{n_k}}\right).$ $[A_{14}]$

3. **Compute the Shapiro-Wilk statistic $SW(i, n_k)$.**
   If H_0 is rejected,
     set $k \leftarrow k + 1$ and go to 1;
     else return n_k.

Let $A_{H_0,\alpha}$ be the acceptance region associated with acceptance level α; this algorithm returns

$$\mathcal{T}_i = \inf_{k \geq 1}\{n_k : SW(i, n_k) \in A_{H_0,\alpha}\},$$

the first time in the sequence of n_k's for which the hypothesis has been accepted. We may in addition plot $SW(i, \cdot)$ and monitor its stabilization in $A_{H_0,\alpha}$. (This will be illustrated in §5.3.4 and for the benchmark examples in §5.5.)

In practice, we need to assess normality of the r.v.'s $N_n(i)$ for several states, e.g. in a subset $E' \subset E$, and Steps 2. and 3. of $[A_{14}]$ can be easily modified to simultaneously test all states $i \in E'$ over the same m simulated sequences. The normality control method then returns $(\mathcal{T}_i, i \in E')$. Automated diagnostics resulting from empirical stopping rules (without graphical monitoring) can be proposed. For instance, a simple rule is to run the chains for at least $\mathcal{T}_{NC} = \max\{\mathcal{T}_i, i \in E'\}$ iterations (the "Normality Control time"). Alternative stopping rules can be considered. The individual level α (i.e. the Type I error for the test) needs to be tuned in such a way that performing $I' = \text{Card}(E')$ tests at this same level results in a preassigned overall level α'. These I' tests are obviously not independent. However, we can apply a technique similar to the Bonferroni's multiple-comparison test for the analysis of variance. This procedure relies on the inequality $\alpha' \leq I'\alpha$ and simply proposes to perform I' individual tests at the asymptotic level α'/I' to obtain an overall level around α'. Hence, the individual level α is tuned by considering the number of states under control and the desired overall level. Finally, the choice of E' (and I') is crucial here, and obviously depends to some extent on the size I of the state space. Chauveau and Diebolt (1997) have proposed empirical procedures for both the discrete and continuous cases. We will generalize their work in §5.4 by proposing a more definitive and automated solution, which encompasses both the discrete case with large I and the continuous state setting.

5.3.3   Convergence assessment by variance comparison

A convergence control tool naturally coupled with the normality monitoring consists in checking whether an estimate of the variance after n steps, $\sigma_n^2(h)$, is close to an estimate of the limiting variance $\sigma^2(h)$. For a single state $i \in E$ (i.e. for $h \equiv \mathbb{I}_i$), the natural estimate of $\sigma_n^2(h)$ based on m parallel chains observed up to the nth transition is simply the sample empirical variance over the m chains,

$$\hat{\sigma}_n^2(m, h) = \frac{1}{nm} \sum_{\ell=1}^{m} \left(N_n^{(\ell)}(i) - \bar{N}_n(i) \right)^2,$$

where

$$\bar{N}_n(i) = \frac{1}{m} \sum_{\ell=1}^{m} N_n^{(\ell)}(i).$$

Besides, an estimate of $\sigma^2(h)$ is available by replacing in (5.9), (5.11) and (5.12), the unknown $\mathbb{P}$, $\mathbb{Z}$, C and π with consistent estimates based on the nm available simulated steps. A natural estimate for the (j, k)-th entry of $\mathbb{P}$ is then given by

$$\hat{\mathbb{P}}_{jk}(m, n) = \frac{\dfrac{1}{m} \sum_{\ell=1}^{m} \sum_{t=1}^{n-1} \mathbb{I}(x_\ell^{(t)} = j, x_\ell^{(t+1)} = k)}{\dfrac{1}{m} \sum_{\ell=1}^{m} \sum_{t=1}^{n-1} \mathbb{I}(x_\ell^{(t)} = j)}. \tag{5.15}$$

A related estimate of π can be obtained from the empirical mean occupation times after nm steps,

$$\hat{\pi}_i(m, n) = \frac{1}{m} \sum_{\ell=1}^{m} \sum_{t=1}^{n} \frac{\mathbb{I}(x_\ell^{(t)} = i)}{n} = \frac{\bar{N}_n(i)}{n}, \quad 1 \leq i \leq I. \tag{5.16}$$

Then

$$\hat{\mathbb{Z}}(m, n) = \{I - [\hat{\mathbb{P}}(m, n) - \hat{A}(m, n)]\}^{-1}, \tag{5.17}$$

(5.11) gives $\hat{C}(m, n)$, and $\hat{\sigma}^2(m, n, h) = h^T \hat{C}(m, n)h$. In this control setup, we are interested in the asymptotic properties of these estimators when n is fixed and m goes to infinity.

Proposition 5.3.1 *For any initial distribution $\tilde{\pi}$ and any fixed integer n large enough, we have*

(i) $\hat{\mathbb{P}}_{jk}(m, n) \to \mathbb{P}_{jk},$

(ii) $\hat{\sigma}^2(m, n, h) \to \sigma^2(h),$

(iii) $\hat{\sigma}_n^2(m, h) \to \sigma_n^2(h),$

a.s. as $m \to +\infty$.

Proof. These results follow from the Strong Law of Large Numbers, with n large enough, typically to allow any state to be reached from any initial state in less than n steps. To illustrate this, consider the strongly aperiodic case, for which $\mathbb{P}_{jk} > 0$ for any j and k. Then running $m \to \infty$ chains for just $n = 1$ step is sufficient for the consistency of $\hat{\mathbb{P}}_{jk}(m, n)$. Generally, *(i)* is proved for fixed $n \geq 2$ since

$$\frac{1}{m} \sum_{\ell=1}^{m} \sum_{t=1}^{n-1} \mathbb{I}(x_\ell^{(t)} = j, x_\ell^{(t+1)} = k) \to \sum_{t=1}^{n-1} \mathbb{P}_{\tilde{\pi}}\left[x^{(t)} = j, x^{(t+1)} = k\right]$$

a.s. as $m \to \infty$, and

$$\sum_{t=1}^{n-1} \mathbb{P}_{\tilde{\pi}}\left[x^{(t)} = j, x^{(t+1)} = k\right] = \sum_{t=1}^{n-1} \mathbb{P}\left[x^{(t+1)} = k \mid x^{(t)} = j\right] \mathbb{P}_{\tilde{\pi}}\left[x^{(t)} = j\right]$$

$$= \mathbb{P}_{jk} \sum_{t=1}^{n-1} \mathbb{P}_{\tilde{\pi}}\left[x^{(t)} = j\right].$$

Similarly,

$$\frac{1}{m} \sum_{\ell=1}^{m} \sum_{t=1}^{n-1} \mathbb{I}(x_\ell^{(t)} = j) \overset{\text{a.s.}}{\underset{m \to \infty}{\to}} \sum_{t=1}^{n-1} \mathbb{P}_{\tilde{\pi}}\left[x^{(t)} = j\right],$$

thus $\hat{\mathbb{P}}_{jk}(m, n) \to \mathbb{P}_{jk}$ a.s. as $m \to \infty$. Using (5.16) we have also that

$$\hat{\pi}_i(m, n) = \frac{1}{n}\left(\frac{1}{m}\sum_{\ell=1}^{m} N_n^{(\ell)}(i)\right) \overset{\text{a.s.}}{\underset{m \to \infty}{\to}} \pi_i, \quad 1 \leq i \leq I,$$

hence $\hat{\pi}(m, n) \to \pi$ a.s. as $m \to \infty$. Using (5.17) and (5.11) gives *(ii)*. The consistency of $\hat{\sigma}_n^2(m, h)$ follows from the Strong Law of Large Numbers applied to the iid random variables $\left(N_n^{(\ell)}(i)\right)^2, 1 \leq \ell \leq m$.   □□

The control by variance comparison uses the setup already described for the normality control. Actually, the two methods can be executed simultaneously, and the original algorithm $[A_{14}]$ becomes:

1. Run the m chains for $(n_k - n_{k-1})$ more iterations.

2. Update the sample $\left(\dfrac{N_{n_k}^{(1)}(i)}{\sqrt{n_k}}, \ldots, \dfrac{N_{n_k}^{(m)}(i)}{\sqrt{n_k}} \right)$. $\qquad\qquad [A_{15}]$

3. Compute the estimates $\hat{\sigma}_{n_k}^2(m, h)$ and $\hat{\sigma}^2(m, n_k, h)$.

4. Compute the Shapiro-Wilk statistic $SW(i, n_k)$.
   If H_0 is rejected,
        set $k \leftarrow k + 1$ and go to 1;
        else return n_k.

In addition to the stopping rule $\mathcal{T}_{NC}$ issued from the normality control, we end up with plots of $\hat{\sigma}_n^2(m, h)$ and $\hat{\sigma}^2(m, n, h)$ against n from which we may check for stabilization and approximate coincidence of the two variance estimators. Note that widely available software systems with algebraic capabilities (e.g., Mathematica, Matlab) can be used to solve the inversion of (5.17), which is hidden in Step 3., without additional work.

These control methods need some adaptation when I gets large. The monitoring using variances comparison requires the computation of the limiting variance, which may not be feasible for large dimension matrices. In such cases this side of the method reduces to graphical monitoring of the stabilization of $\hat{\sigma}_n^2(m, h)$, with no guarantee against apparent stabilization far from the limiting variance. This would result in wrong convergence diagnostics and biased confidence intervals. This is the reason why this control method should not be used alone, but rather coupled with the normality control. The latter is less affected by the size of the chains from a computational perspective (no matrix is involved in the computations), but it can lead to a dramatically conservative method for large I's. Moreover, estimating the probabilities π_i for the I states $i \in E$ is meaningless and misleading when I is large (just think of continuous state spaces). A reasonable way to overcome this problem is to choose a partition $(A_1, \ldots, A_p)$ of E and to apply the normality control method to the corresponding indicator functions $h_1 = \mathbb{I}_{A_1}, \ldots, h_p = \mathbb{I}_{A_p}$. In that case, we can only obtain estimations and confidence intervals for the probabilities $\pi(A_1), \ldots, \pi(A_p)$. Also, this partitioning suggests another implementation of the control by variance comparison, in the spirit of Raftery and Lewis' (1992a) binary method described in §2.2.2. These adapted versions are very similar to the proposed extensions for the continuous state case, and we thus postpone their complete description till §5.4.

5.3.4   A finite example

To illustrate the behaviour of the two control methods involved in algorithm $[A_{15}]$ we show, on a toy example, how the stabilization processes of the variance estimates and test statistics differ between two finite chains: one known to mix quickly and the other leading to a multimodal setting. We selected two chains $(x^{(t)})$ and $(y^{(t)})$, with $I = 15$ states. The transition matrix for x is a normalized matrix drawn at random. The chain allows for quick transitions between all states, and the resulting invariant probability $\pi(x)$ (Figure 5.1, *left*) is roughly uniform. For y (Figure 5.1, *right*), we artificially built a matrix generating a multimodal invariant probability $\pi(y)$ with three "disconnected" modal regions $(1, 2, 3)$, $(7, 8, 9)$ and the smaller mode $(13, 14, 15)$. For both examples, we ran $m = 50$ independent parallel chains, started according to a uniform initial distribution over $E = \{1, \ldots, 15\}$. We then controlled these chains with $[A_{15}]$ using the asymptotic level $\alpha = 0.01$. Since we simultaneously controlled four to six states here, this value of α roughly results in an overall asymptotic level around 0.05. In addition to the stopping rule $\mathcal{T}_{NC}$ of §5.3.2, we implemented a more conservative rule: *"stop at the first time when all the controlled states simultaneously do not reject the null hypothesis"*. The estimated invariant probabilities in Figure 5.1, and Student's t 95%–confidence intervals based on the normality assumption for the sample of occupation times were computed at the latter stopping time, which was not much larger than $\mathcal{T}_{NC}$ in this example. We always ran the parallel chains up to a fixed large amount of iterations to show stabilization after the stopping time.

FIGURE 5.1. Invariant probabilities $\pi(x)$ *(left)* and $\pi(y)$ *(right)*. The true probabilities are in gray, and their estimates at convergence time in black.

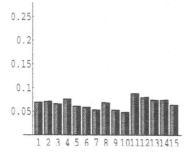 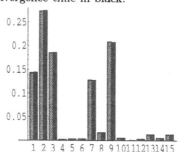

For the simple chain x, convergence occurred after only 100 to 200 iterations, depending on the four arbitrarily selected controlled states. Both stopping rules gave the same results. All the confidence intervals based on the normality assumption at this time contained the true value. Figure 5.2 *(left)* shows a typical output, for state 2 for which we obtain the 95%-confidence interval $[0.060, 0.074]$ for the true value $\pi_2(x) = 0.0706$ after 200 iterations. Note that the time required to assess normality, $\mathcal{T}_2$, is small here, and we just ran the chains up to 10,000 iterations. The stabilization of $\hat{\sigma}_n^2(m, \mathbf{I}_2)$ seems to be achieved after about 2400 iterations, and shows merely noise around $\hat{\sigma}^2(m, n, \mathbf{I}_2)$ after that. The latter estimate stabilizes in a time comparable to $\mathcal{T}_2$.

For the "multimodal" chain y, we applied our controls over one state near each mode and one state between two contiguous modal regions. The first important observation is that convergence (in the normality control sense and for both stopping rules) occurred after about 100,000 iterations, depending on the selected states (101,000 iterations in this example where states $\{2, 5, 8, 11, 15\}$ were controlled). Again, at convergence, the Student's t 95%–confidence interval contained the true value for each of the controlled states, and we ran the chains up to 200,000 iterations to show stabilization. State 2 (not displayed here) is in the most frequently visited modal region. Thus stabilization is achieved comparatively quickly (about 10,000 iterations according to the normality check). Controls for states in the second modal region exhibited the same stabilization time. Surprisingly enough the same behaviour held for state 5, although $\pi_5(y) = 0.00298$ is very small. Finally, the normality check provided the most conservative time for states 13, 14 and 15. Although $\pi_{15}(y)$ is four times larger than $\pi_5(y)$, the normality control required 101,000 iterations, and the Shapiro-Wilk statistic stabilized after that. We just display the control for state 15 here (Figure 5.2, *right*); other states (except 13 and 14) are very similar to the control displayed in Figure 5.2 *(left)*. This typical behaviour for the chain y arises because for this chain many jumps occur between the first two modes. Therefore, the chain frequently visits states $4, 5, 6$, leading rather quickly to a normal-like histogram for, e.g., state 5. When started outside the third mode, the chain usually does not visit it for a very long time. Roughly 1/5 of the 50 chains started from this third mode and spent time within it, whereas the remaining 4/5 started outside this mode and remained outside it for many iterations. This explains why samples of occupation times for state 15 are far from normality at the beginning. In this example, about 100,000 iterations allow for enough visits to this third mode for most of the 50 chains to attain normality. This miniature example shows that it is basically multimodality, and not only estimation of small probabilities for the stationary distribution (e.g., $\pi_5(y)$), that really affects the normality of samples of occupation times. It also highlights the ability of a parallel chain method to detect such delicate situations.

5.4   Extension to continuous state chains

Consider an ergodic Markov chain $\left(x^{(t)}\right)$ with continuous state space E and invariant probability density f. Suppose in addition that $\left(x^{(t)}\right)$ satisfies

FIGURE 5.2.   *Left*: chain x, control for state 2. The true variance is $\sigma^2(\mathbb{I}_2) = 0.0598$. The normality control gives $\mathcal{T}_2 = 100$. *Right*: chain y, control for state 15. The true variance is $\sigma^2(\mathbb{I}_{15}) = 2.494$. The normality control gives $\mathcal{T}_{15} = 101,000$.

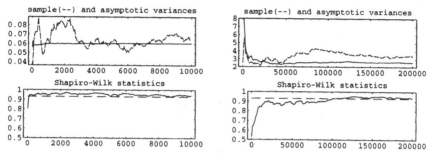

conditions ensuring that the CLT applies, i.e. that for every $h \in L_2(f)$, there exists $0 \leq \sigma^2(h) < +\infty$ such that

$$\frac{1}{\sqrt{n}} \sum_{t=1}^{n} \left(h(x^{(t)}) - \mathbb{E}^f[h] \right) \overset{\mathcal{L}}{\leadsto} \mathcal{N}\left(0, \sigma^2(h)\right). \tag{5.18}$$

Various sets of sufficient conditions for the CLT in the context of general MCMC's have been stated in §5.2.3.

We do not base our extension to the continuous case on renewal theory for atoms or small sets. Since the construction of appropriate small sets requires some knowledge relative to the transition kernel or the target density f (see Chapter 4 and the benchmark examples in §4.5), this would result in strongly problem-specific control techniques(thus not in the spirit of this normality control principle). Rather, we suggest to select a finite collection of measurable subsets $A_r \subset E$, $1 \leq r \leq p$, typically "almost" partitioning E, and to check normality and variance stabilization of the normalized sums $S_n(h_r)/\sqrt{n}$, where $h_r = \mathbb{I}_{A_r}$ for $1 \leq r \leq p$.

Similarly to the finite case with large I's, we can only obtain estimates and confidence intervals for the $\mathbb{P}^f(A_r)$'s; moreover, since an estimate of the limiting variance can no longer be algebraically computed with formulas (5.9) to (5.12), another control of the variance stabilization must be done.

5.4.1   *Automated normality control for continuous chains*

This approach through a partition of E is theoretically valid in the continuous setup and in particular does not require any Markovian assumption on the $h_r(x^{(t)})$'s. Furthermore, it can apply to general processes $(x^{(t)})$, provided that they converge to a unique stationary regime and satisfy a Strong Law of Large Numbers and a CLT similar to (5.18). This is of particular interest since in general marginal sequences issued from multivariate MCMC algorithms are *not* Markov chains. In addition, approximate normality of $S_n(h)/\sqrt{n}$ for particular functions h may be checked simultaneously. For instance, we have always tested the approximate normality for $h(x) = x$ (or higher moments) in the illustrative examples, since posterior means for the parameters are generally desired in Bayesian setups. In multivariate situations, marginal sequences were controlled in the same way.

We propose a methodology for monitoring a general MCMC algorithm, which does not require prior knowledge on the target pdf f, and consequently which has been included into our generic computer code for the normality control. In classical settings where the support E of f is the real line or an infinite denumerable set, we obviously cannot measure the tails of f over E accurately: tail regions with almost zero probability would require a dramatically large number of iterations to reach approximate normality, without a noticeable improvement in the precision over estimates like (1.2). Therefore, a preliminary requirement is to restrict our investigations to a suitable compact connected subset $\mathcal{A}$ of E, which needs to be chosen large enough so that $\mathbb{P}^f(\mathcal{A})$ is close to one. The normality hypothesis may be checked for a collection of indicator functions h_r of subsets A_r of equal length, such that $\mathcal{A} = \bigcup_{r=1}^p A_r$. The "controlled region" $\mathcal{A}$ and the "sharpness" p of this partition of $\mathcal{A}$ are preliminary parameters of the procedure. The controlled region $\mathcal{A}$ needs to be chosen fairly large, and this choice (without preliminary knowledge of f) has to be validated by the estimate $\hat{\mathbb{P}}(\mathcal{A})$ given on-line by the algorithm. The sharpness p is directly related to the final desired precision for the approximate picture of f given by the histogram $(\hat{\mathbb{P}}(A_1), \ldots, \hat{\mathbb{P}}(A_p))$. Then, the idea is to perform the normality control only over the normalized sums of indicator functions of the A_r's representing a significant probability, e.g. such that $\hat{\mathbb{P}}(A_r) > \varepsilon$ for a tuning parameter ε, where these estimated probabilities are updated on-line along with the parallel simulations. The regions representing a non significant proportion of the total mass are thus simply discarded from the set of controlled functions. These regions would typically correspond to tails of f, or to regions between two almost disconnected modes.

More formally, let $C(n)$ be the set of indicator functions $h_r = \mathbb{I}_{A_r}$ which correspond to subsets of significant estimated probabilities for which the normality hypothesis has not yet been accepted at time n, with $C(0) = \{h_1, \ldots, h_p\}$. The sequence of sets $C(n)$ is decreasing since, at time n, normalized indicator functions that have reached approximate normality or which correspond to subsets of too small probabilities are deleted from $C(n)$. A validation of this deletion procedure is given on-line by the algorithm, in term of the estimated mass of the subset $\mathcal{A}_C \subset \mathcal{A}$ consisting of the *controlled* sets at convergence,

$$\hat{\mathbb{P}}(\mathcal{A}_C) = \sum_{r=1}^p \mathbb{I}_C(A_r)\, \hat{\mathbb{P}}(A_r), \tag{5.19}$$

where $\mathbb{I}_C(A_r) = 1$ if A_r has been controlled and finally accepted, and $\mathbb{I}_C(A_r) = 0$ if A_r has been discarded.

For m parallel chains $x_1^{(t)}, \ldots, x_m^{(t)}$ started from an initial distribution $\tilde{\pi}$ uniform over $\mathcal{A}$ (or even over a larger subset of E), we define the sum $S_n(h)$ for the chain ℓ by

$$S_n^{(\ell)}(h) = \sum_{t=1}^{n} h(x_\ell^{(t)}), \quad \ell = 1, \ldots, m,$$

and the consistent estimate of $\mathbb{P}(A_r)$ over the parallel chains by

$$\hat{\mathbb{P}}(A_r) = \frac{\bar{S}_n(h_r)}{n}, \quad \text{where } \bar{S}_n(h_r) = \frac{1}{m} \sum_{\ell=1}^{m} S_n^{(\ell)}(h_r). \tag{5.20}$$

Given arbitrary increasing times $n_0 = 0 < n_1 < n_2 < \cdots$, the algorithm for controlling the posterior distribution is:

0. **Start from step** $k = 1$.

1. **Run the** m **chains for** $(n_k - n_{k-1})$ **more iterations.**

2. **For** $r = 1, \ldots, p$ **update the samples**

$$\left(\frac{S_{n_k}^{(1)}(h_r)}{\sqrt{n_k}}, \ldots, \frac{S_{n_k}^{(m)}(h_r)}{\sqrt{n_k}} \right).$$

3. **For** $r = 1, \ldots, p$ **update** $\hat{\mathbb{P}}(A_r)$; $[A_{16}]$
   update $C(n_k) = \left\{ h_r \in C(n_{k-1}) : \hat{\mathbb{P}}(A_r) \geq \varepsilon(n_k) \right\}$.

4. **For each** $h \in C(n_k)$:
   compute the statistics $SW(h, n_k)$;
   if H_0 **is accepted for** $SW(h, n_k)$, $C(n_k) \leftarrow C(n_k) \setminus \{h\}$.

5. **If** $C(n_k) = \emptyset$, **return** n_k;
   else set $k \leftarrow k + 1$ **and go to 1.**

The sequence $\varepsilon(n)$ which appears in $[A_{16}]$ is just a refinement of the probability threshold ε. Its purpose is to lower the effect of the poor estimations of the $\mathbb{P}^f(A_r)$'s that we may obtain during the first few iterations of the m chains. Actually, we do not want to wrongly discard an A_r from being controlled, just because of an under-estimation of $\mathbb{P}^f(A_r)$. Choosing a sequence $\varepsilon(n)$ increasing smoothly from 0 to ε may avoid such a behaviour.

After K steps, this algorithm returns a "normality control time" $\mathcal{T}_{NC}$ which corresponds, as in the discrete case, to the largest number of iterations n_K required to reach approximate normality in all the subsets A_r

with significant estimated probability. For each $h_r \in C(0)$ controlled up to acceptance of H_0, the time to reach approximate normality is

$$\mathcal{T}_r = \min_{k \geq 1} \{n_k : SW(h_r, n_k) \in A_{H_0, \alpha}\}$$

and $\mathcal{T}_{NC} = n_K$ is the maximum of these times. To be meaningful, this result must be validated by the estimated mass of the region on which the control has been imposed, $\hat{\mathbb{P}}(\mathcal{A})$, and the estimated mass of the sets within which approximate normality has been reached, $\hat{\mathbb{P}}(\mathcal{A}_C)$ given by (5.19). Both probabilities should be close to one. Too small a value for $\hat{\mathbb{P}}(\mathcal{A})$ indicates a wrong choice for $\mathcal{A}$, which leaves a significant proportion of the total mass outside of it. Too small a value for $\hat{\mathbb{P}}(\mathcal{A}_C)$ indicates that a significant mass has not been controlled (typically in the tails or between distant modes), and consequently that ε needs to be lowered down.

The algorithm $[A_{16}]$ ends up with the overall normality control time, and a detailed picture of f exhibiting all its specificities (e.g., modal regions), together with precise estimates and confidence intervals for the $\mathbb{P}^J(A_r)$'s, based on reliable normal approximations. Simultaneously, normality control over the mean of the parameter or its higher posterior moments are provided. They may be used to give a more conservative stopping rule, and to compute confidence intervals for the parameter (or marginal coordinates in multivariate situations) using the approximate normality.

5.4.2   Variance comparison

In the continuous setup, we can still consistently estimate the variance after n steps, $\sigma_n^2(h)$, by the sample empirical variance

$$\hat{\sigma}_n^2(m, h) = \frac{1}{nm} \sum_{\ell=1}^{m} \left(S_n^{(\ell)}(h) - \bar{S}_n(h) \right)^2, \tag{5.21}$$

where $\bar{S}_n(h)$ is given in (5.20). Unfortunately, the algebraic computations leading to an estimate for $\sigma^2(h)$ are no longer feasible and we need some sort of discretization of the continuous Markov chain to mimic the discrete case. Since we do want to keep the generic aspect of this methodology intact, we are not deriving our discretization from small sets as in Chapter 4. Instead, we propose to apply a discretization directly over the partition $(A_1, \ldots, A_p, A_{p+1})$ of E (where $A_{p+1} = E \setminus A$), by considering the process

$$\xi^{(t)} = \sum_{r=1}^{p+1} r \mathbb{I}_{A_r}(x^{(t)}) \tag{5.22}$$

which takes values in $\{1, \ldots, p+1\}$. This can be seen as a generalization of the binary discretization proposed by Raftery and Lewis (1992a) in their control method, and is not any more valid from a theoretical perspective, since $(\xi^{(t)})$ is not a Markov chain for two reasons.

1. When $(x^{(t)})$ is a Markov chain, $(\xi^{(t)})$ does not (usually) satisfy the *lumpability* condition (see Kemeny and Snell, 1960).

2. In multivariate situations, the marginal $(x^{(t)})$ is not even a Markov chain.

We could determine a *batch size* (number of iterations ignored between two recordings of the Markov chain) as in Raftery and Lewis (1992a), but this seems to make the implementation more difficult without bringing actual improvement, as discussed in §2.2.2. We thus propose this approximation as a trade-off between theoretically-valid discretization and easily implementable and generic control method.

If we consider the process $\xi^{(t)}$ as a discrete Markov chain over the state space $\{1, \ldots, p+1\}$, with pseudo-transition matrix $\mathbb{P}_\xi$, and related matrices $\mathbb{Z}_\xi$, A_ξ and C_ξ as in §5.3.3, we can algebraically derive, for each $r \in \{1, \ldots, p\}$, an estimate $\hat{\sigma}_\xi(m, n, \mathbb{I}_r)$ for the limiting variance $\sigma_\xi^2(\mathbb{I}_r)$ which can be compared with the estimate of $\sigma_n^2(h_r)$ given by (5.21) for $h_r = \mathbb{I}_{A_r}$, which is computed on the continuous chain $(x^{(t)})$. As in the discrete case, this estimate may serve as a graphical tool to check whether the variance after n steps stabilizes around a value which may be considered here as an approximation of the true limiting variance. From the implementation point of view, $[A_{16}]$ needs only to be augmented to compute the above estimates along with the tests for the normality hypothesis, for each $h_r \in C(n_k)$ at each step k.

5.4.3 A continuous state space example

We consider the two-component normal mixture distribution

$$p\mathcal{N}(\mu_1, \sigma_1) + (1-p)\mathcal{N}(\mu_2, \sigma_2) \tag{5.23}$$

which has been presented in §3.4, with the same conjugate priors, and use the Gibbs implementation given by $[A_9]$, which simulates iteratively the missing data and the parameter $\theta = (p, \mu_1, \sigma_1^2, \mu_2, \sigma_2^2)$. Given a sample of size 30 from (5.23) using the true parameter $\theta^* = (0.3, -3, 1, 3, 4)$, we applied[2] our control method $[A_{16}]$ together with the variance comparison in the following way:

1. marginally for each scalar parameter, control over the posterior marginal with the *automated partition method* of §5.4.1;

2. marginally, control over each scalar coordinate;

---

[2] From an implementation point of view, the generic code for our control tools only requires to be linked with the code producing the next step given the current state for this multivariate Gibbs sampler.

3. for the multivariate parameter, *global* control over the scalar function

$$h_G(\theta) = p + \mu_1 + \sigma_1^2 + \mu_2 + \sigma_2^2.$$

The selection of the controlled region $\mathcal{A}$ for each coordinate was easily done by a short run of $[A_{16}]$ for a few iterations, which sketched out the mass location for each marginal posterior. The resulting estimates $\hat{\mathbb{P}}(\mathcal{A})$ were always larger than 99.9%. The threshold was set to $\varepsilon = 0.004$, resulting in estimates for $\hat{\mathbb{P}}(\mathcal{A}_C)$ between 98.5% and 99.3%. Convergence in our sense, and for all the controlled functions, always occurred before $n = 2000$ iterations, and we ran the $m = 50$ parallel chains up to $10,000$ iterations to show stabilization.

Figure 5.3 shows a set of results for two coordinates. The values for the $(\hat{\mathbb{P}}(A_1), \ldots, \hat{\mathbb{P}}(A_p))$ at time $\mathcal{T}_{NC}$ are represented by the histograms together with their confidence intervals using the normal approximation. To save space, we just give for each coordinate control plots for the $h_r = \mathbb{I}_{A_r}$ requiring the largest time to reach approximate normality (hence in this situation typically in the tails). Note that the approximate estimates for the limiting variance $\hat{\sigma}_\xi(m, n, h_r)$, available just for the indicator functions, behave as in the discrete case: they stabilize rather quickly, but not always around the average value of the sample variance. This may be a side-effect of the discretization, or an effect of the long-memory which characterizes the sample variance process. However, as in the discrete case, they may be used as a complementary tool coupled with the normality control. In this example, the function $h_G(\theta)$ which considers the multivariate parameter, and hence can be seen as a *global* control for this MCMC algorithm, required 1600 iterations to reach approximate normality. Actually, the largest convergence time was obtained for the parameter σ_1^2 (i.e. for the control over $h(\theta) = \sigma_1^2$) which required 2000 iterations. On the other side, the indicator functions of the A_r's located near the modes of the posterior marginals stabilized in less than 100 iterations, as expected.

5.5   Illustration for the benchmark examples

5.5.1   Cauchy Benchmark

We consider here the posterior distribution (1.6) of Example 1.2.3, and the Gibbs sampler given in Example 2.2.1. This example is one-dimensional, and our control methods directly applies. The posterior is multimodal, with three modes located around the observations $x_1 = -8$, $x_2 = 8$ and $x_3 = 17$.

The selection of the controlled region $\mathcal{A}$ was done, as in previous examples, by a short run of $[A_{16}]$ for a few iterations, which sketched out the location of the mass and resulted in the estimate $\hat{\mathbb{P}}(\mathcal{A}) = 99.7\%$. We imposed a sharpness $p = 50$ here, to get a reasonable precision for the trimodal

FIGURE 5.3. Control for μ_1 (*left*), and σ_2^2 (*right*). Each column gives successively the estimated marginal posterior distributions at convergence time with Student's t confidence intervals (*in black*), the control for selected A_r's in the tails of the posterior, and the control for the coordinates. In the bottom the control for $h_G(\theta)$ is given. Each control consists in two plots: the variance comparison with the approximate asymptotic variance in dashed lines when available (*top*), and the Shapiro-Wilk statistic, with its acceptance region above the horizontal dashed line (*bottom*).

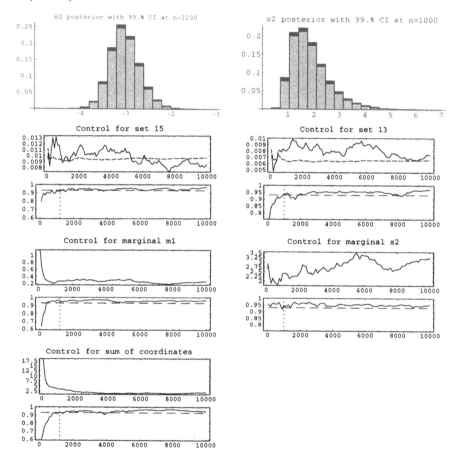

histogram of the posterior distribution, and the threshold for controlling the sets A_r's was set to $\varepsilon = 0.002$ to gain control over $\hat{\mathbb{P}}(\mathcal{A_C}) = 99.0\%$ of total mass. Note that p and ε are linked somehow, since the probabilities $\mathbb{P}^f(A_r)$ decrease when p increases, and consequently, the times needed to reach approximate normality increase (mostly for the A_r's in the tails or located between modes, where less and less jumps are observed when the A_r's decrease). For this reason, p should not be larger than a value imposed by the precision wanted for the "picture" of f. In other words, the more precision we want for the histogram of f, the more time it takes to get this picture with approximate normality. In addition to our control over the posterior distribution, we controlled the functions $h_1(\theta) = \theta$ and $h_2(\theta) = \theta^2$. It is interesting to point out that the additional stopping rules associated with the control of h_1 and h_2 are not influenced by the choices made for p and ε, and in this sense act as moderators.

Approximate normality occurred around $n = 3400$ iterations, and we ran the $m = 50$ parallel chains up to $10,000$ iterations to show stabilization. Figure 5.4 shows some selected results. As expected, the normality control time $\mathcal{T}_{NC}$ returned by $[A_{16}]$ corresponds to a set A_{15} located around -2, between the largest mode and the smallest distant mode. Note that the Shapiro-Wilk statistic for this set (Figure 5.4, *right*) required much more time than the "slow sets" in the mixture example (Figure 5.3) to stabilize in the acceptance region. The plots for h_1 and h_2 show a quick stabilization; normality is reached after less than 500 iterations. This indicates that the region of small probability between the two distant modes, although requiring 3400 iterations to reach approximate normality, has little influence over the estimation of the parameter θ, or of θ^2. This can be linked with the quick mixing behaviour of the chain which requires on average less than 20 iterations to visit the small mode around -8.

5.5.2   Multinomial Benchmark

We consider the continuous chain over the two-dimensional parameter $\theta = (\mu, \eta)$. The selection of the control region $\mathcal{A} = [0; 1]$ for each marginal was straightforward here, and resulted in $\hat{\mathbb{P}}(\mathcal{A}) = 1$. We set $p = 20$ and $\varepsilon = 0.004$, which provided enough precision for these unimodal marginal posterior distributions. We controlled the posteriors, the marginals, and the functions $h_1(\theta) = \mu + \eta$ and $h_2(\theta) = \mu^2 + \eta^2$. The normality control times returned by $[A_{16}]$ and essentially given in Figure 5.5, were between 50 (parameter marginal) and 1100 ("slow" set A_r), indicating a fast convergence for this simple chain. The validation of the control was insured by the estimates $\hat{\mathbb{P}}(\mathcal{A}_c) = 99.2\%$ for μ and 99.7% for η. Note that increasing p to 50 (the sharpness used for the Cauchy benchmark) just increased the convergence time for μ from 600 to 1100, and kept the convergence time of 1100 for η, for which $[A_{16}]$ just discarded some additional sets of too small probabilities, lowering $\mathbb{P}(\mathcal{A}_c)$ to 98.6%. Another run with $\varepsilon = 0.002$ brought $\hat{\mathbb{P}}(\mathcal{A}_c)$ back to 99.5%, and just increased the convergence time to 1500. Hence, this chain converges actually faster than the chain used in the Cauchy benchmark, and the observed difference is not a side effect of the tuning parameters.

FIGURE 5.4. Control for the Cauchy benchmark. Estimated posterior distribution at convergence time with Student's t confidence intervals (*in black*). In the second row, control for A_r's corresponding to the fastest (*left*) and slowest (*right*) convergence times. In the third row, control for the parameter and for $h_2(\theta) = \theta^2$. Each control consists in two plots: the variance comparison with the approximate asymptotic variance in dashed lines when available (*top*), and the Shapiro-Wilk statistic, with its acceptance region above the dashed line (*bottom*).

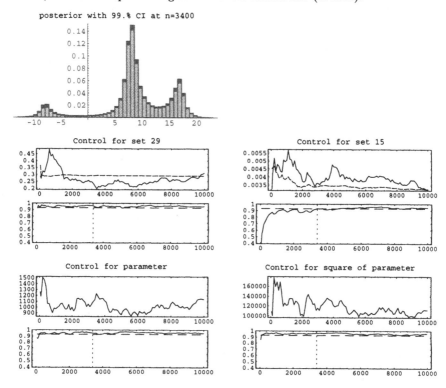

5.6   Testing normality on the latent variables

In most set-ups,

$$S_T(h) = \frac{1}{\sqrt{T\,\sigma^2(h)}} \sum_{t=1}^{T} \left(h(x^{(t)}) - \mathbb{E}^f[h(x)] \right) , \qquad (5.24)$$

with

$$\sigma^2(h) = \text{var}^f\left(h(x^{(0)})\right) + 2 \sum_{t=1}^{\infty} \text{cov}^f\left(h(x^{(0)}), h(x^{(t)})\right) ,$$

is thus asymptotically distributed as a $\mathcal{N}(0,1)$ random variable. In addition to the previous developments, which take specifically advantage of the

FIGURE 5.5. Control for the multinomial benchmark: μ (*left*), and η (*right*). Each column gives successively the estimated marginal posterior distributions at convergence time with Student's t confidence intervals (*in black*), the control for selected (slow) A_r's in the tails of the posterior, and the control for the coordinates. Last row gives the control for $h_1(\theta) = \mu + \eta$ (*left*) and $h_2(\theta) = \mu^2 + \eta^2$ (*right*). Each control consists in two plots: the variance comparison with the approximate asymptotic variance in dashed lines when available (*top*), and the Shapiro-Wilk statistic, with its acceptance region above the horizontal dashed line (*bottom*).

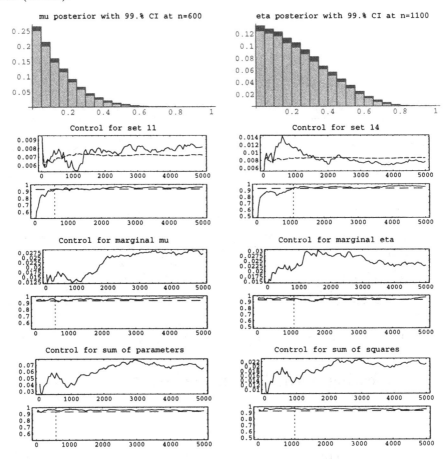

finite nature of the Markov chain or of its discretization, we present in this section an alternative control procedure based on the CLT, and introduced in Robert, Rydén and Titterington (1998). The main point of including this method in the book is that, while the CLT obtained by Robert *et al.* (1998) applies in full generality, it takes its full meaning as a control technique in latent variable models, in particular when the latent variables have finite supports.

As in the previous sections (§5.3 and §5.4), we try to assess whether or not the normality of (5.24) holds, in order (a) to evaluate if stationarity is attained and (b) to propose confidence regions on the quantities of interest $\mathbb{E}^f[h(x)]$. A major difference with the above method is that Robert *et al.* (1998) use the CLT on (5.24) to test the convergence of a *single* sequence $(x^{(t)})$, instead of using *parallel* replications of the same chain. Given the previous developments and criticisms, in particular on the *"You've only seen where you've been"* defect of single chain evaluations, this choice seems rather paradoxical. In addition, (5.24) relies on two unknown quantities, $\mathbb{E}^f[h(x)]$ and $\sigma^2(h)$, of which one is the quantity of interest! The second quantity, $\sigma^2(h)$, is particularly difficult to evaluate,[3] because of the Markovian structure of the chain. It is in fact equivalent to estimating the spectral density function of $h(x^{(t)})$ at frequency zero (see, e.g., Geyer, 1992, or Geweke, 1992).

The idea behind the method of Robert *et al.* (1998) is to somehow relinquish some of the information contained in the chain by using a subsampling device, as in §4.3, on the chain $(x^{(t)})$ with the additional constraint that the sampling intervals grow with time. Heuristically, this means that the dependence between subsamples will asymptotically vanish, and that the limiting variance corresponding to $\sigma^2(h)$ will simply be $\text{var}^f(h(x))$. Obviously, subsampling also implies that we introduce some randomness in the control method, but this is a necessary counterpart to the simplicity of the method.

More precisely, given $x^{(1)}, \ldots, x^{(T)}, \ldots$, output of an MCMC algorithm, Robert *et al.* (1998) use the standard estimators of $\mathbb{E}^f[h(x)]$ and $\text{var}^f(h(x))$,

$$\hat{\mu}_T = T^{-1} \sum_{t=1}^{T} h(x^{(t)}), \qquad \hat{\sigma}(h)_T^2 = T^{-1} \sum_{t=1}^{T} h^2(x^{(t)}) - (\hat{\mu}_T)^2,$$

which are strongly convergent. (We here assume that h is a scalar function, but the extension to the multivariate case is straightforward.) Subsampling times (or *instants*) t_k are then generated by

$$t_{k+1} - t_k - 1 \sim \mathcal{P}oi(\nu k^d), \tag{5.25}$$

where $\nu \geq 1$ and $d > 0$, so that the difference $t_{k+1} - t_k$ is increasing with k on average. If $N_t = \sup\{n : t_n \leq t\}$ denotes the number of such sampling instants before time t, we can consider the normalized sum based on the subsampled chain,

$$\tilde{S}_T = \frac{1}{\sqrt{N_T \, \hat{\sigma}_T^2(h)}} \sum_{k=1}^{N_T} \left\{ h(x^{(t_k)}) - \hat{\mu}_T \right\}. \tag{5.26}$$

---

[3]Except when the setting of §4.6.1 applies.

In order to derive a useful CLT on the S_T's, we need additional assumptions on the mixing behaviour of the chain $(x^{(t)})$ (see Bradley, 1986, for details).

Definition 5.6.1 *A chain* $(x^{(t)})$ *is said to be* strongly mixing *(or α-mixing)* *if*

$$\alpha(k) = \sup_{n \geq 0} \sup_{A \in \mathcal{F}_1^n, B \in \mathcal{F}_{n+k}^\infty} |P(A \cap B) - P(A)P(B)| \downarrow 0$$

as $k \to \infty$, *where* $\mathcal{F}_1^n$ *and* $\mathcal{F}_{n+k}^\infty$ *are the σ-fields generated by* $(x^{(1)}, \ldots, x^{(n)})$ *and* $(x^{(n+k)}, x^{(n+k+1)}, \ldots)$, *respectively. If* $\alpha(k) \leq C\beta^k$ *for some* $C \geq 0$ *and* $\beta \in (0, 1)$, *the mixing coefficients are said to be* geometrically decaying.

Robert, Rydén and Titterington (1998) then establish the following CLT:

Theorem 5.6.1 *If the chain* $(x^{(t)})$ *is ergodic and strongly mixing with geometrically decaying mixing coefficients, and if*

$$\mathbb{E}^f |h(x)|^{2+\delta} < \infty, \qquad \delta > 0,$$

then $\tilde{S}_T$ *converges weakly to the standard normal distribution.*

The condition of geometrical decay is rather mild (see Meyn and Tweedie, 1993, Chapter 16). In fact, every positive recurrent aperiodic Markov chain is strongly mixing (Rosenblatt, 1971). Note also that the Lyapunov condition on $h(x)$ is a sufficient condition for the regular CLT to hold on ϱ-mixing (Rosenblatt, 1971). As stressed by Robert *et al.* (1998), if the mixing coefficients of $(x^{(t)})$ decay slower than geometrically, the conclusion of Theorem 2.1 still holds if ν_k grows sufficiently fast and/or if the Poisson distribution is replaced by a different one.

The main feature of this CLT is thus that the limiting distribution of $\tilde{S}_T$ is the standard normal $\mathcal{N}(0, 1)$ distribution and does not involve complex quantities like $\sigma(h)$. Moreover, the usual estimates $\hat{\mu}_T$ and $\hat{V}_T$ can be replaced by other convergent estimates such as Rao-Blackwellized versions (see §3.2). It is thus bound to provide a more manageable basis for control methods than the standard CLT of §5.2. The fact that the method requires subsampling can, however, be perceived as a drawback since it discards some of the information contained in the sample and thus may force more iterations than necessary. But this is not the case for the estimates $\hat{\mu}_T$ and $\hat{\sigma}(h)_T^2$, which take advantage of the whole sample. Note also that the method can be implemented on-line and does not require a modification of the original MCMC algorithm.

In the setup of latent-variables models such as mixture models, hidden Markov models, or state space models, this result can be transformed into a convergence control device in a natural way. Indeed, the latent variables z_j, like the allocations in mixture models (see §3.4) or in hidden Markov chains, are generally generated at each iteration of the MCMC algorithm.

For observations x_j $(j = 1, \ldots, n)$, it is thus possible to subsample (independently in j) from $(z_j^{(t)})$, while deriving the estimates $\widehat{\mu}_{jT}$ and $\widehat{V}_{jT}$ (or their Rao-Blackwellized counterparts) from the whole chain. If N_j denotes the number of subsampled $z_j^{(t)}$'s at time T, with corresponding sampling times t_{jk} $(1 \le k \le N_j)$, we then derive normalized versions of the average latent variables

$$\zeta_j = N_j^{-1/2} \widehat{\sigma}_j^{-1} \sum_{k=1}^{N_j} (z_j^{(t_{jk})} - \widehat{\mu}_j) \,, \qquad (5.27)$$

for $j = 1, \ldots, n$, which should be approximately iid $\mathcal{N}(0, 1)$ for T large enough. As mentioned in §1.5, another advantage of this setting is that geometric α-mixing holds naturally, given that the z_j's have finite support (Billingsley, 1968). For each time T, the normality of the resulting sample $\zeta_1, \ldots, \zeta_n$ can therefore be tested through a one-sample Kolmogorov-Smirnov test or a Shapiro-Wilk test as in the previous sections. (In practice, the test is run every 100 or 1000 iterations.)

Getting back to the example of §3.4, we are thus endowed with a simple evaluation of the convergence through the normality of the sample $(\zeta_1, \ldots, \zeta_n)$, the $z_j^{(t)}$'s being generated from Step 1. of $[A_{10}]$. In this case, the mean and variance of the $z_i^{(t)}$'s can be evaluated through Rao–Blackwellisation, namely by estimating the stationary distribution of $z_j^{(t)}$, a $\mathcal{M}_k(\pi_1, \ldots, \pi_k)$ distribution, through the Rao-Blackwellized estimates

$$\hat{\pi}_j \propto \frac{1}{T} \sum_{t=1}^{T} \frac{p_j^{(t)}}{\sigma_j^{(t)}} \exp\left\{ -(x_i - \theta_j^{(t)})^2 / 2(\sigma_j^{(t)})^2 \right\} .$$

FIGURE 5.6. Normal fit of the allocation average when standardized with respect to the Rao-Blackwellized estimates of the allocation probabilities, incl. histogram of the standardized averages, normal pdf and QQ-plot.

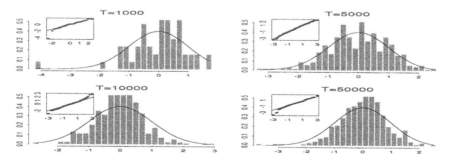

Figure 5.6 illustrates the result of this assessment for different values of T, by providing the histogram of the standardized averages ζ_j which are

defined (i.e. for which the asymptotic standard deviation $\hat{\tau}_i^{(T)}$ is not equal to 0) and the corresponding QQ-plot. Note the progressive alignement in the QQ-plots and the corresponding improvement in the normal approximation of the histogram. (See Chapters 6, 7 and 8 for other illustrations.)

6
Convergence Assessment in Latent Variable Models: DNA Applications

Florence Muri
Didier Chauveau
Dominique Cellier

6.1 Introduction

A DNA sequence is a long succession of four *nucleotides* or *bases*, Adenine, Cytosine, Guanine and Thymine, and can be represented by a finite series $x = (x_1, \cdots, x_n)$, each base x_t taken from the alphabet $\mathcal{X} = \{A, C, G, T\}$. It turns out that there is an important heterogeneity within the genome.[1] Statistical models based on a complete homogeneity assumption are thus unrealistic. We propose a hidden Markov chain approach to identify homogeneous regions in the DNA sequence. The breakpoints which define these regions may thus separate parts of the genome with different functional or structural properties.

In this model (introduced in §6.2), one assumes that the DNA sequence has a mosaic structure composed of homogeneous regions and that there is a finite number k of models providing a good description of each region. The sequence of regions is described by an unobservable k-state Markov chain, the hidden state chain $z = (z_1, \cdots, z_n)$. The aim is to reconstruct these regions from the DNA sequence and to estimate the parameter θ of the k models in order to characterize the identified regions.

Churchill (1989) has used the EM algorithm (introduced by Dempster, Laird and Rubin, 1977, for incomplete data) to compute the maximum likelihood estimate of such a hidden Markov model and to identify homogeneous regions in DNA sequences (see for instance Rabiner, 1989, Qian and Titterington, 1990, 1991, Celeux and Clairambault, 1992, Juang and Rabiner, 1991, Archer and Titterington, 1995, for a review of identification procedures of hidden Markov chains). To avoid some well-known drawbacks of the EM algorithm (such as convergence to local maxima), we propose a Bayesian estimation using the Gibbs sampler described in §6.2.2 (see Muri

---

[1] The entire DNA content of an organism is called its *genome*.

1997, 1998). In this setup, the Gibbs sampler generates two chains in parallel: a continuous one, the parameter chain $(\theta^{(t)})$, and a discrete one, the hidden state chain $(z^{(t)})$. Thus, the Duality Principle (see §1.5) allows us to apply convergence control methods on both chains.

Section 6.3 presents the results obtained for the $bIL67$ bacteriophage, including the estimation of the parameters and the hidden states (§6.3.1), and the classical convergence diagnostics (§6.3.2) using the CODA software of Best *et al.* (1995). Section 6.4 uses the perfect sampling method (CFTP) of Propp and Wilson (1996) (see §1.4) to start the hidden state chain $(z^{(t)})$ in its stationary regime. The last section is devoted to the control by the CLT as in Chapter 5: §6.5.1 presents the results for the parameter θ using parallel chains as in Chauveau and Diebolt's method in §5.4, and §6.5.2 uses the method of Robert *et al.* (1998) to test the normality of the latent variables $(z^{(t)})$.

6.2   Hidden Markov model and associated Gibbs sampler

6.2.1   M1-M0 hidden Markov model

Hidden Markov models are characterized by two processes (see for instance Rabiner, 1989): the hidden states process $z = (z_1, \cdots, z_n)$, such that $z_i \in \mathcal{Z} = \{0, \cdots, k-1\}$ (which in our setup governs the arrangement of the k possible regions along the sequence), and the observed process $x = (x_1, \cdots, x_n)$, $x_i \in \mathcal{X} = \{A, C, G, T\}$, corresponding to the observed DNA sequence. The states are generated according to an homogeneous first order Markov chain with transitions $(2 \leq i \leq n)$

$$a_{uv} = P(z_i = v \mid z_{i-1} = u), \quad 0 \leq u, v \leq k-1, \quad \text{with} \quad \sum_{v=0}^{k-1} a_{uv} = 1 \, ,$$

and with initial distribution

$$a_v = P(z_1 = v), \quad 0 \leq v \leq k-1, \quad \sum_{v=0}^{k-1} a_v = 1 \, ,$$

assumed to be the stationary distribution.

The bases appear in the sequence with a distribution which depends on the hidden states. The *M1-M0* model assumes that, conditionally on the state z_i, the bases are drawn independently with probability $(1 \leq i \leq n)$

$$b_{va} = P(x_i = a \mid z_i = v), \ 0 \leq v \leq k-1, \ a \in \mathcal{X}, \quad \text{with} \quad \sum_{a \in \mathcal{X}} b_{va} = 1 \, .$$

Hence, this model takes into account the composition of the bases in the sequence and corresponds to the classical hidden Markov model described in the literature. More generally, the *M1-Mm* model assumes an *m*-order Markovian dependence between the observations conditionally on the hidden states and allows us to account for the local structure in oligonucleotides of length *m* of the DNA sequence (see Churchill, 1989, and Muri, 1998).

We apply this modelling to the *bIL67* bacteriophage: this phage consists of 22,195 basepairs[2] (*bp*) and is a parasite of the *Lactococcus lactis* bacterium. We only present the results obtained for a *M1-M0* model with $k = 2$ hidden states.[3] The parameter of the model is denoted by

$$\theta = (a_{00}, a_{00}, b_{0A}, b_{0G}, b_{0C}, b_{1A}, b_{1G}, b_{1C})$$

and belongs to a 8-dimensional space Θ.

Hidden Markov chains are thus missing data models, more exactly mixture models with dependent data. The aim is to reconstruct the hidden states to identify homogeneous regions in the *bIL67* genome and to estimate θ to characterize the identified regions.

Let $f(x \mid \theta)$ be the likelihood of the incomplete data x and $g(x, z \mid \theta)$ the likelihood of the complete data (x, z) related to f by

$$f(x \mid \theta) = \sum_{z \in \{0,1\}^n} g(x, z \mid \theta) = \sum_{z \in \{0,1\}^n} \left(a_{z_1} b_{z_1 x_1} \prod_{i=2}^{n} a_{z_{i-1} z_i} b_{z_i x_i} \right) . \quad (6.1)$$

The computation of (6.1) involves a sum of 2^n terms and is thus intractable for large values of n. Hence, the Bayesian estimation of θ is difficult to perform from the incomplete model. As for the mixture setup, the solution is to use an MCMC approximation to calculate the posterior distribution $\pi(\theta \mid x)$, based on the complete model $g(x, z \mid \theta)$.

In the following, the notation $y_{i_1}^{i_2}$ refers to the group of $i_2 - i_1 + 1$ consecutive elements $(y_{i_1}, y_{i_1+1}, \cdots, y_{i_2-1}, y_{i_2})$.

6.2.2   MCMC implementation

As in Robert, Celeux and Diebolt (1993), we consider independent Beta priors $\mathcal{Be}(1, 1)$ for each row $a(u) = (a_{u0}, a_{u1})$ of the state transition matrix

---

[2] DNA has a double helicoidal structure composed of two strands which are antiparallel in orientation and complementary (an A on one strand is paired with a T on the other strand, and a G is always paired with a C). Thus the complete information is contained in the sequence of one strand.

[3] The number k of hidden states could be determined by using the reversible jump MCMC of Richardson and Green (1997) or the criterion of Mengersen and Robert (1996), based on the Kullback-Leibler distance.

and Dirichlet priors $\mathcal{D}(1,\ldots,1)$ for the rows $b(u) = (b_{uA}, b_{uG}, b_{uC}, b_{uT})$ of the observation probability matrix $(u = 0, 1)$.

We use a *Data Augmentation* type scheme to generate iteratively the hidden states and the parameters. The states are simulated from their joint conditional distribution $\pi(z \mid x, \theta)$, derived from the relation

$$P(z_1, \cdots, z_n \mid x, \theta) = P(z_n \mid x, \theta) \ldots P(z_i \mid z_{i+1}^n, x, \theta) \ldots P(z_1 \mid z_2^n, x, \theta) ,$$

and θ is simulated from the conditional posterior distribution

$$\pi(\theta \mid z, x) \propto \pi(\theta) g(x, z \mid \theta) .$$

The two steps of the Gibbs sampler are:

1. Simulate the missing data z by $(1 \le i \le n)$ $[A_{17}]$

$$z_i \sim \mathcal{M}\left(1, P(z_i = 0 \mid z_{i+1}^n, x, \theta), P(z_i = 1 \mid z_{i+1}^n, x, \theta)\right) ,$$

where the probabilities $P(z_i = u \mid z_{i+1}^n, x, \theta)$, $u = 0, 1$, are calculated by the following *forward-backward recurrence* introduced by Baum, Petrie, Soules and Weiss (1970) to compute the EM estimate in the hidden Markov chain setup (see also, e.g., Rabiner, 1989, or Qian and Titterington, 1990).

Forward recurrence

1. Take $P(z_1 = u \mid x_1^1, \theta) = a_u$ $(u = 0, 1)$.

2. For $i = 2, \cdots, n$, compute $(u = 0, 1)$

$$P(z_i = u \mid x_1^i, \theta) \propto b_{vx_i} \sum_{u=0}^{1} a_{uv} P(z_{i-1} = u \mid x_1^{i-1}, \theta) .$$

Backward recurrence

1. Deduce $P(z_n = u \mid x, \theta) = P(z_n = u \mid x_1^n, \theta)$ $(u = 0, 1)$.

2. For $i = n - 1, \cdots, 1$, compute $(u = 0, 1)$

$$P(z_i = u \mid z_{i+1}^n, x, \theta) \propto a_{uz_{i+1}} P(z_i = u \mid x_1^i, \theta) .$$

The second step of the Gibbs iteration is then

2. Simulate each row $(u = 0, 1)$ $[A_{18}]$

$$a(u) \sim \mathcal{B}e\left(1 + \sum_{i=2}^{n} \mathbf{I}_{\{z_{i-1}=u, z_i=0\}}, 1 + \sum_{i=2}^{n} \mathbf{I}_{\{z_{i-1}=u, z_i=1\}}\right) ,$$

$$b(u) \sim \mathcal{D}\left(1 + \sum_{i=1}^{n} \mathbf{I}_{\{z_i=u, x_i=A\}}, \cdots, 1 + \sum_{i=1}^{n} \mathbf{I}_{\{z_i=u, x_i=T\}}\right) .$$

As pointed out in §3.4, the global state allocation $[A_{17}]$ is time-consuming. To avoid the forward-backward recurrence on the sequence length, an alternative is to use a local state allocation derived from the Gibbs decomposition $[A_{11}]$ (see for instance Robert *et al.*, 1993, Muri, 1998 and Robert, Rydén and Titterington, 1998) by considering the full conditional distributions

$$P(z_i = u \mid \cdots, z_{i-1}, z_{i+1}, \cdots, \theta, x) \propto a_{z_{i-1}z_i} b_{z_i x_i} a_{z_i z_{i+1}},$$

while θ is still generated as in $[A_{18}]$. Simulation results (see Muri, 1997) show that both allocation schemes give similar results for large numbers of iterations, but that the local allocation becomes stable more slowly than the global one, especially when the DNA sequence is composed of a few large homogeneous regions (and this will be the case for $bIL67$, see §6.3).

Simulating from $[A_{17}]$ and $[A_{18}]$ generates two dual Markov chains: $(z^{(t)})$ is an irreducible aperiodic Markov chain with finite state space $\tilde{Z} = \{0, 1\}^n$. The Duality Principle (see §1.5) applies and states that the continuous Markov chain $(\theta^{(t)})$ has the same properties as the hidden state chain $(z^{(t)})$. From a theoretical point of view, both chains are then uniformly geometrically ergodic and the Central Limit Theorem applies (note that with a local state allocation, $(\theta^{(t)})$ is not a Markov chain but the Duality Principle still holds, as shown by Robert *et al.*, 1993).

We will present in the next sections convergence diagnostic results to monitor the estimated posterior mean $T^{-1} \sum_{t=1}^{T} \theta^{(t)}$ and, for each position $i = 1, \cdots, n$ in the sequence, the estimated state probabilities

$$\hat{P}(z_i = u) = \frac{1}{T} \sum_{t=1}^{T} \mathbb{I}_{\{z_i^{(t)} = u\}}, \quad u = 0, 1.$$

6.3   Analysis of the $bIL67$ bacteriophage genome: first convergence diagnostics

6.3.1   Estimation results

First, we present the results obtained with a single run of 10,000 iterations of the Gibbs sampler initialized with the following parameter value $\theta^{(0)} = (0.999, 0.998, 0.6, 0.1, 0.1, 0.1, 0.2, 0.3)$. Such a choice corresponds to initial ranges of 1000 bp mean size for the first region ($a_{00}^{(0)} = 0.999$) alternating regularly with 500 bp mean size ranges for the second region; moreover, the compositions for both regions are quite contrasted.

The analysis of a 2 hidden state model reveals two homogeneous regions with little uncertainty (as shown by Figure 6.1): the $bIL67$ genome is clearly separated in two. Table 6.1 shows that the first region is rich in A's and poor in C's, whereas the second region has a high content of T's and a low content of G's. From a biological point of view, these two regions should correspond to different transcription directions of the $bIL67$ genes: the first region should contain the genes located on one of the DNA strands (see Muri, 1997) and the second region the genes located on the other strand.

FIGURE 6.1. Estimated probabilities $\hat{P}(z_i = 1)$ of the states z_i based on 10,000 iterations, against the sequence position i: identification of two homogeneous regions for the $bIL67$ bacteriophage. (*Source:* Muri, 1997)

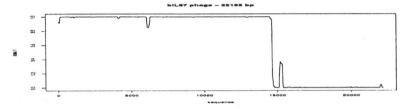

TABLE 6.1. MCMC estimates and batch mean standard errors (batch size of 25) for the 8 parameters of the *M1-M0* model, based on a single run of 10,000 iterations (with no iteration discarded).

parameter	estimate	batch SE
a_{00}	0.99979	0.00002
a_{11}	0.99980	0.00001
b_{0A}	0.3952	0.0001
b_{0G}	0.19164	0.00005
b_{0C}	0.14030	0.00004
b_{1A}	0.27095	0.00007
b_{1G}	0.16124	0.00003
b_{1C}	0.21298	0.00004

6.3.2 Assessing convergence with CODA

The **CODA** software (Best, Cowles and Vines, 1995) allows us to apply classical convergence diagnostics to the parameter chain $(\theta^{(t)})$. First, we present the results obtained on the single run of 10,000 iterations. We then use the between-within variance criterion of Gelman and Rubin (1992) based on 30 parallel runs.

A first analysis of the output chain produced by **CODA** is a graphical summary of the 10,000 iterations for each parameter. Figure 6.2 does not show any particular problem with convergence, although this rudimentary graphical method does not allow to state convergence. On the other hand, the parameters autocorrelations (computed with all the iterations), presented in Figure 6.3, are reasonable on the 10,000 iterations, even though it clearly detects a slower convergence of the parameters a_{00} and a_{11} (note that, with a thinning interval of 10 iterations, the autocorrelations are almost 0 from lag 5 for all the parameters).

FIGURE 6.2. Raw plots of the 8 parameter chains, obtained by CODA.

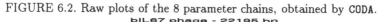

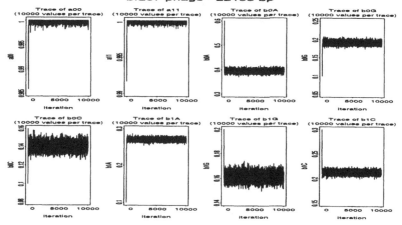

FIGURE 6.3. Autocorrelations for the 8 parameters, based on 10,000 iterations.

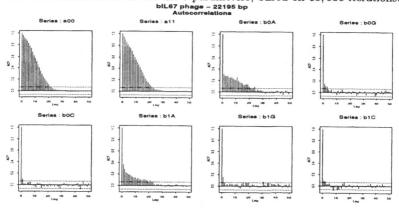

We then apply the following convergence evaluations to this single output chain:

(a) the binary approximation of Raftery and Lewis (1992a,b) described in §2.2.2, with control parameters $q = 0.025$, $\epsilon = 0.0005$ and $\epsilon' = 0.95$;

(b) the normality test of the Z-scores of Geweke (1992), based on the first 10% and the last 50% iterations;

(c) the stationarity test of Heidelberger and Welch (1983).

We briefly recall the basis of the diagnostics (b) and (c), as in Mengersen, Robert and Guihenneuc-Jouyaux (1998). Following a remark in Hastings (1970), the chain $(\theta^{(t)})$ or a transform $(h(\theta^{(t)}))$ can be considered as a *time series*. Geweke (1992) uses the *spectral density* of $h(\theta^{(t)})$, $S_h(w)$, which gives the asymptotic variance $\gamma_h^2 = S_h^2(0)$ of the empirical average of the $h(\theta^{(t)})$'s. The spectral density can be estimated by nonparametric methods. Geweke (1992) then takes the first T_A observations and the last T_B observations out of T iterations and computes the partial expectations

$$\delta_A = \frac{1}{T_A} \sum_{t=1}^{T_A} h(\theta^{(t)}), \quad \delta_B = \frac{1}{T_B} \sum_{t=T-T_B+1}^{T} h(\theta^{(t)}) ,$$

as well as the corresponding estimates σ_A^2 and σ_B^2 of $S_h^2(0)$. The difference (Z-score),

$$\frac{\sqrt{T}(\delta_A - \delta_B)}{\sqrt{\dfrac{\sigma_A^2}{\tau_A} + \dfrac{\sigma_B^2}{\tau_B}}} ,$$

is (asymptotically in T) a standard normal variable (with $T_A = \tau_A T$ and $T_B = \tau_B T$). This asymptotic normality then induces a convergence diagnostic as well as a determination of the burn-in time t_0. As suggested by Geweke (1992), we have chosen $\tau_A = 0.1$ and $\tau_B = 0.5$.

Another approach based on spectral analysis is Heidelberger and Welch's (1983), which uses Cramér–von Mises procedures to test stationarity (see also Schruben, Singh and Tierney, 1983). The diagnostic is based on the statistic

$$B_T(S) = \frac{S_{[Ts]} - [Ts]\bar{\theta}}{(T\hat{\psi}(0))^{1/2}} , \qquad 0 \leq s \leq 1 ,$$

where

$$S_t = \sum_{t=1}^{t} \theta^{(t)}, \qquad \bar{\theta} = \frac{1}{T} \sum_{t=1}^{T} \theta^{(t)} ,$$

$[a]$ is the integer part of a, and $\hat{\psi}(0)$ is an estimate of the spectral density at 0. For large T's, B_T is approximately a Brownian bridge, hence inducing a possible test.

The conclusions produced by CODA are the following: Raftery and Lewis' (1992) evaluation suggests a very optimistic warm-up time (see Table 6.2) and states that 10,000 iterations seem to be more than enough to estimate b_{0A}, b_{0G}, b_{0C}, b_{1G} and b_{1C}, if not for a_{00}, a_{11} and b_{1A}. The conclusions of the diagnostics of Geweke (1992) (see Figure 6.4) and Heidelberger and Welch (1983) (Table 6.3) are more or less the same: they give a positive signal for b_{0G} and b_{1C} while they suggest to discard the first 1000 iterations for the other parameters. Note that all these methods point out a lack of convergence for a_{00}, a_{11} and b_{1A}.

TABLE 6.2. Raftery and Lewis' (1992) convergence diagnostic for the 8 parameters obtained by CODA, with control parameters $q = 0.025$, $\epsilon = 0.0005$ and $\epsilon' = 0.95$ (based on 10,000 iterations).

parameter	t_0	T
a_{00}	20	24,975
a_{11}	28	28,794
b_{0A}	2	3942
b_{0G}	2	3819
b_{0C}	2	3696
b_{1A}	16	17,768
b_{1G}	2	3802
b_{1C}	2	3802

TABLE 6.3. Heidelberger and Welch's (1983) stationarity test for the 8 parameters obtained by CODA (based on 10,000 iterations).

parameter	stationarity test	# iters to discard	# iters to keep
a_{00}	*passed*	1000	9000
a_{11}	*passed*	1000	9000
b_{0A}	*passed*	1000	9000
b_{0G}	*passed*	0	10,000
b_{0C}	*passed*	1000	9000
b_{1A}	*passed*	1000	9000
b_{1G}	*passed*	1000	9000
b_{1C}	*passed*	0	10,000

Finally, we present the results of Gelman and Rubin's (1992) evaluation (see §2.3.2) based on 30 parallel runs of 10,000 iterations. The starting point $\theta^{(0)}$ of each run was generated as follows: the persistence state probabilities a_{00} and a_{11} were chosen uniformly in $[0.8, 1]$ while the observation probabilities b_{ua}, $u = 0, 1$ and $a \in \mathcal{X}$, were all uniformly generated in $[0.1, 0.6]$. CODA produces the 50% and 97% shrink factors estimated from the second half of each run. Gelman and Rubin's (1992) factors are equal to 1 for all parameters: this suggests that 5000 iterations are sufficient to achieve convergence and that the last 5000 iterations may be

FIGURE 6.4. Geweke's (1992) diagnostic plot for the 8 parameters based on 10,000 iterations, obtained by CODA.

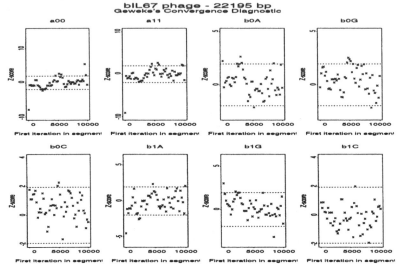

FIGURE 6.5. Gelman and Rubin's (1992) diagnostic plot for the 8 parameters based on 30 runs of 10,000 iterations, obtained by CODA.

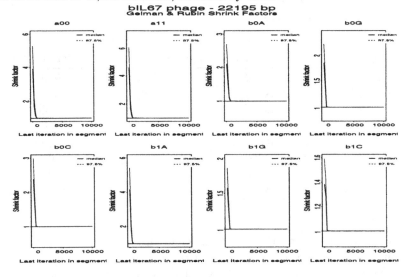

assumed to arise from the marginal posterior distribution for each parameter. Figure 6.5 illustrates the evolution of both factors and shows that both quantiles stabilize around 1 for all parameters after roughly the first 500 iterations; this tends to prove that stationarity was first achieved after approximately 250 iterations.

6.4 Coupling from the past for the *M1-M0* model

In this section, we use the perfect sampling method (CFTP) of Propp and Wilson (1996), described in §1.4, to start the hidden state chain $(z^{(t)})$ in its stationary regime.

We only present the case of two hidden states ($k = 2$) in the DNA *M1-M0* model introduced in §6.2.1. The method can be generalized to any k hidden state *M1-Mm* model (see Muri, 1997).

6.4.1 The CFTP method

Here, the finite hidden state space is $\tilde{\mathcal{Z}} = \{0, 1\}^n$. The CFTP method couples chains with all possible starting points in $\tilde{\mathcal{Z}}$ farther and farther back in time until all chains coalesce at time 0, as described in § 1.4.

Recall that the states and the parameters are simulated from $[A_{17}]$ and $[A_{18}]$ respectively. The transition of the hidden state chain is then given by

$$p_{zz'} = P\left(z^{(t+1)} = z' \mid z^{(t)} = z\right) = \int_\Theta \pi(\theta \mid z, x)\pi(z' \mid \theta, x)d\theta$$

$$= \int_\Theta \pi(\theta \mid z, x) \prod_{i=1}^n \pi(z_i' \mid z_{i+1}'^n, \theta, x)d\theta .$$

We use one iteration of the Gibbs sampler to generate z' according to this transition:

1. **Given** z, **simulate** $\theta_z \sim \pi(\theta \mid z, x)$.

2. **Let** $w = (w_1, \cdots, w_n)$, $w_i \sim \mathcal{U}_{[0,1]}$, **generate** $z' = (z_1', \cdots, z_n')$ **by**

$$z_i' = \phi_i(z, w_i) = \begin{cases} 0 & \text{if} & w_i < \pi(0 \mid z_{i+1}'^n, \theta_z, x) \\ 1 & \text{if} & w_i \geq \pi(0 \mid z_{i+1}'^n, \theta_z, x) \end{cases} .$$

For all $t > 0$, the applications $f_{-1}, f_{-2}, \cdots, f_{-t}, \cdots$ from $\tilde{\mathcal{Z}}$ to $\tilde{\mathcal{Z}}$, constructed by the CFTP method (see (1.7)) are then deduced from a single vector of uniform variables $w^{(t)}$ by

$$\forall z \in \tilde{\mathcal{Z}}, \qquad f_{-t}(z) = \phi(z, w^{(t)}) ,$$

where $\phi = (\phi_1, \phi_2, \cdots, \phi_n)$. Hence, the functions $F_{-t} = F_{-t+1} \circ f_{-t}$, with $F_0 = id_{\tilde{\mathcal{Z}}}$, describe the paths of all the states of $\tilde{\mathcal{Z}}$ from time $-t$ until present time 0.

The CFTP is stopped when F_{-N} is constant. The time N of coalescence is dynamically determined by the algorithm, and the random variable $F_{-N} = z^{(0)}$ produced by the CFTP method is distributed from the stationary distribution of the chain $(z^{(t)})$. Note that at every time t, the coupling method is not independent because we use the same uniform variable for every starting point in $\tilde{\mathcal{Z}}$. In addition, we have to keep all the uniform variables and this is expensive in space requirement.

6.4.2 The monotone CFTP method for the M1-M0 DNA model

Since the hidden state space $\tilde{\mathcal{Z}}$ have 2^n points, it is impossible, for time and storage reasons, to make 2^n couplings at each time t. However, the problem can be solved because a *stochastically monotone structure* is available on $\tilde{\mathcal{Z}}$.

We can define a partial order on $\tilde{\mathcal{Z}}$, deduced from the natural order $0 \leq 1$ by

$$z \leq z' \iff \forall i = 1, 2, \cdots, n, \quad z_i \leq z_i' \ .$$

The space $\tilde{\mathcal{Z}}$ has then a smaller state $\tilde{0} = (0, 0, \cdots, 0)$ and a larger one $\tilde{1} = (1, 1, \cdots, 1)$.

Generally, we suppose that the homogeneous regions are sufficiently large, so that the transition of the hidden Markov chain is close to the identity (see the results presented on $bIL67$ phage in the previous section). Therefore, the conditional distributions $\pi(z_i \mid z_{i+1}^n, \theta, x)$ are attractive: the function $z \to \pi(1 \mid z_{i+1}^n, \theta, x)$ is a nondecreasing function of z (recall that we only impose a partial order on $\tilde{\mathcal{Z}}$).

If $z \leq \tilde{z}$, for all $i = 1, 2, \cdots, n$ and $v = 0, 1$ we have

$$\sum_{u=1}^{v} \pi(u \mid z_{i+1}^n, \theta, x) \geq \sum_{u=1}^{v} \pi(u \mid \tilde{z}_{i+1}^n, \theta_z, x) \ .$$

Hence, if, at every time t, we use the same uniform variables $w^{(t)}$ for all the paths,

$$\phi(z, \cdot) \leq \phi(\tilde{z}, \cdot) \ ,$$

and the order on $\tilde{\mathcal{Z}}$ is preserved by the CFTP (note that when we use a local allocation for the states, the CFTP technique still preserves the order, see Muri, 1997).

In this case, it is sufficient to consider two chains starting in states $\tilde{0}$ and $\tilde{1}$; all the intermediary paths are located between the two extreme cases. When $F_{-N}(\tilde{0}) = F_{-N}(\tilde{1})$, the function F_{-N} is constant. Finally, Propp and Wilson (1996) suggest to use an overestimation of the coalescence time N which gives good results (see also Murdoch and Green, 1998). We try successively $N = 1, 2, 4, \cdots 2^t, \cdots$ until coalescence.

We can summarize the monotone CFTP method as follows:

1. **Take** $N = 1$.

2. **Repeat** $[A_{19}]$

(a) initialize $\sup = \tilde{1}$ and $\inf = \tilde{0}$.

(b) for $t = -N, -N+1, \cdots, -1$ generate with the same vector of uniform variables $w^{(t)}$.

$$
\begin{aligned}
\sup &= \phi(\sup, w^{(t)}) \\
\inf &= \phi(\inf, w^{(t)})
\end{aligned}
$$

(c) Take $N = 2N$.

until $\sup = \inf$.

3. Return $\sup = F_{-N}(\tilde{0}) = F_{-N}(\tilde{1}) = z^{(0)}$, distributed from the stationary distribution.

Note that we use the same $w^{(t)}$ for all time t already run in previous iterations.

6.4.3 Application to the bIL67 bacteriophage

The application of the monotone CFTP algorithm $[A_{19}]$ to $bIL67$ leads to going back in time for only $N = 256$ iterations to reach coalescence. To control the contribution of the CFTP on the parameter and state estimation, we compare the MCMC estimates obtained from the run of 10,000 iterations described in the previous section, and from the run of 10,000 iterations of the Gibbs sampler initialized with the stationary distribution given by the CFTP method.

Although the parameter estimates are similar in both cases (see Table 6.4), Figure 6.6 shows clearly a higher stability of the MCMC estimate along iterations when the Gibbs sampler is initialized with the CFTP method, especially for the persistence state probabilities a_{00} and a_{11}. The estimates seem to be stable almost from the first iterations. Moreover, all the different diagnostics of CODA applied to the run initialized with the CFTP (with the same control parameters as in §6.3.2) give a positive signal: the Geweke's (1992) Z-scores fall within the 95% confidence interval, Raftery and Lewis' (1992a) diagnostic suggests between 2 and 8 iterations for the burn-in time and a convergence time between 3740 (for b_{0C}) and 9780 (for a_{00}) iterations. Heidelberger and Welch's (1983) evaluation declares stationarity for all the parameters (with no iteration discarded) and the autocorrelations are almost 0 from lag 6 (and from lag 1 with a thinning interval of 10 iterations). Note that Geweke's (1992) and Heidelberger and Welch's (1983) assessments are still positive if we only retain the first 1000 iterations. These results tend to prove that the CFTP technique does not only allow to suppress the approximation of the stationary distribution of $(z^{(t)})$ but also to start its dual chain $(\theta^{(t)})$ in a stationary regime, which is a consequence of the Duality Principle.

FIGURE 6.6. Evolution of the MCMC estimates for the 8 parameters on 10,000 iterations of the Gibbs sampler initialized with (*full line*) or without (*dashed line*) CFTP.

blL67 phage – 22195 bp

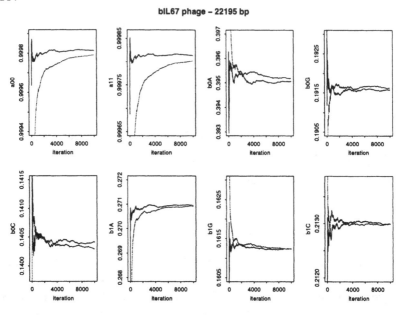

TABLE 6.4. MCMC estimates for the 8 parameters, based on 10,000 iterations, initialized without or with the CFPT technique (with no iteration discarded).

parameter	estimate without CFTP	estimate with CFTP
a_{00}	0.99979	0.99981
a_{11}	0.99980	0.99981
b_{0A}	0.3952	0.3951
b_{0G}	0.1916	0.1916
b_{0C}	0.1403	0.1404
b_{1A}	0.2709	0.2710
b_{1G}	0.1612	0.1612
b_{1C}	0.2130	0.2130

The interest of the CFTP method for the hidden state estimation can be evaluated by comparing the evolution of the states allocation along the iterations. Figure 6.7 presents an *allocation map* (introduced in §3.4 for the mixture example), which gives the successive allocations of the 22,195 states (black for state 0 and grey for state 1) along the 100 first iterations and the 10,000 first iterations of the sampler initialized with and without

FIGURE 6.7. Allocation map of the 22, 195 states along the iterations (*black for state 0 and grey for state 1*). *Upper:* Allocations along 10,000 iterations. *Lower:* Allocations along the first 100 iterations. *Left:* MCMC algorithm initialized without the CFTP technique. *Right:* MCMC algorithm initialized with the CFTP technique.

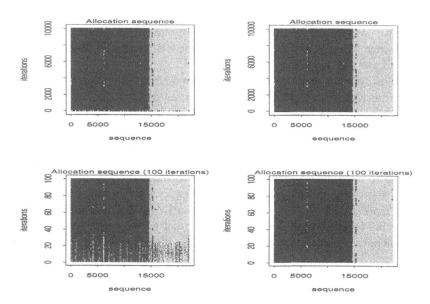

the CFTP technique. The upper graphs show a real and fast stability in the allocations along the 10, 000 iterations whatever the starting point. Note that, as shown in the lower graphs, the allocation of the states seems to become stable from the very first iterations when we use the CFTP method to initialize the Gibbs sampler. These results confirm those obtained for the parameters.

Even if the MCMC estimates (states and parameters) are quite the same in both settings, the two chains $(\theta^{(t)})$ and $(z^{(t)})$ become stable faster when we use Propp and Wilson's (1996) method. However, a real drawback of the CFTP technique is the large number of uniform variables we need to store to check coalescence. In our setup, these methods are thus impractical for very long DNA sequences (since they require the storage of n random variables at each iteration of the CFTP) or for known situations of slower convergence of the Gibbs sampler, like alternation of short regions in the sequence (see Muri, 1997).

6.5   Control by the Central Limit Theorem

6.5.1   Normality control for the parameters with parallel chains

In this section, we apply the normality control of Chauveau and Diebolt, presented in Chapter 5, through the automated partition method described in §5.4.1, with the algorithm $[A_{16}]$. We use again the 30 parallel chains from the Gibbs sampler on which we performed Gelman and Rubin's (1992) control method in §6.3.2.

Following the methodology already presented for the illustrative and the benchmark examples (§5.4.3 and §5.5), we used short runs of $[A_{16}]$ to determine appropriate controlled regions $\mathcal{A}$'s, marginally for each scalar parameter. These choices resulted in estimates $\hat{P}(\mathcal{A})$'s always around 99%. Note that this desired precision required more than 10,000 iterations for the persistence states probabilities (a_{00} and a_{11}) to stabilize properly, hence we ran up to 30,000 iterations for those parameters.

Estimates for the $P(\mathcal{A})$'s, and for the regions into which normality has been reached, $P(\mathcal{A}_C)$'s (where $\mathcal{A}_C$ is given by (5.19)), are summarized in Table 6.5 together with the convergence times and the confidence intervals for the parameters. Graphical results are given in Figures 6.8, 6.9, and 6.10.

TABLE 6.5. Normality control results for 30 parallel chains. The table provides the normality control time $\mathcal{T}_{NC}$, the Student's t confidence intervals (CI) for the parameters at their corresponding convergence time, and the estimated probabilities of $\mathcal{A}$ and $\mathcal{A}_C$ defined in Chapter 5.

parameter	$\mathcal{T}_{NC}$	CI	$\hat{P}(\mathcal{A})$	$\hat{P}(\mathcal{A}_C)$
a_{00}	5500	[0.9980 , 0.9988]	99.03%	99.03%
a_{11}	12,500	[0.9991 , 0.9994]	99.32%	99.32%
b_{0A}	7400	[0.3958 , 0.3962]	99.42%	99.36%
b_{0G}	7000	[0.1911 , 0.1913]	99.89%	99.67%
b_{0C}	9800	[0.1407 , 0.1409]	99.61%	97.93%
b_{1A}	6000	[0.2689 , 0.2698]	98.91%	98.34%
b_{1G}	7800	[0.1614 , 0.1615]	99.81%	99.54%
b_{1C}	5800	[0.2126 , 0.2128]	99.85%	99.62%

It is interesting to point out that for each scalar parameter, approximate normality is reached very quickly. This is in accordance with the diagnostic given by Gelman and Rubin's (1992) control method. However, it takes much more time to reach approximate normality in the tails of the marginal posteriors, particularly when multimodality occurred, as for a_{00} and b_{0C}, or when posteriors have long thin tails, as for b_{0A} or b_{1C}. As noted in Chapter 5, this is quite normal since achieving normality for the posteriors requires more time than assessing stationarity for the parameters. Also, the plots of the sample empirical variances for all the parameters clearly show that several thousands iterations are necessary to achieve the limiting variances stabilization. In comparison, the few hundred iterations diagnosed by Gelman and Rubin's (1992) method seem quite unrealistic.

FIGURE 6.8. Normality control for a_{00} and a_{11}. Each column gives successively the estimated marginal posterior distributions at convergence time with Student's t confidence intervals (in $black$), the control for selected sets in the tails of the posterior, and the control for the coordinates. Each control consists in two plots: the variance comparison with the approximate asymptotic variance in dashed lines when available (top), and the Shapiro-Wilk statistic, with its acceptance region above the horizontal dashed line ($bottom$).

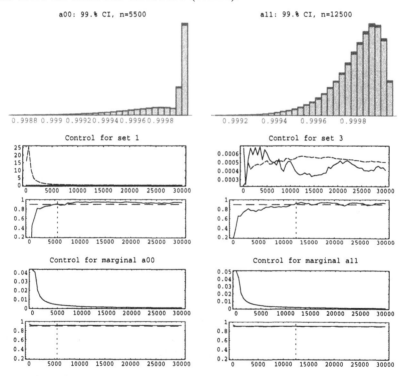

6.5.2   Testing normality of the hidden state chain

In this section, we use the normality test proposed by Robert, Rydén and Titterington (1998) (§5.6) to assess the convergence of the hidden states. Remember that the Markov chain $(z^{(t)})$ has a finite state space and thus the geometric α-mixing condition, required for this method, holds. In a

FIGURE 6.9. Normality control for b_{0A}, b_{0C}, b_{0G}. Same legend as Figure 6.8.

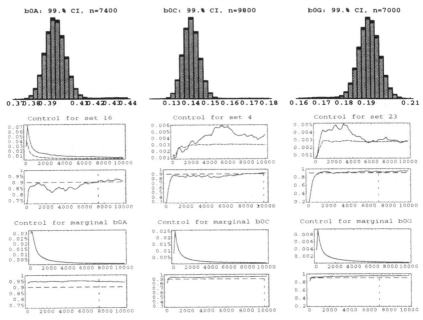

FIGURE 6.10. Normality control for b_{1A}, b_{1C}, b_{1G}. Same legend as Figure 6.8.

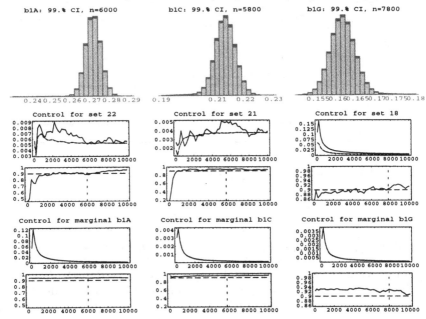

$k = 2$ hidden state *M1-M0* model, each state z_i, $i = 1, \cdots, n$ is generated according to [A_{17}], namely from a Bernoulli variable $\mathcal{B}(\mu_i)$ where $\mu_i = P(z_i = 1 \mid z_{i+1}^n, x, \theta)$. The diagnostic is based on the sample of the normalized sums

$$\frac{1}{\sqrt{N_i \hat{\sigma}_i^2}} \sum_{k=1}^{N_i} \left(z_i^{(t_{ik})} - \hat{\mu}_i \right), \quad i = 1, \cdots, n , \tag{6.2}$$

where the times t_{ik} are subsampling times such that (see §5.6)

$$t_{i\,k+1} - t_{ik} - 1 \sim \mathcal{P}oi(10k^{0.01}) ,$$

where N_i is the number of subsampled $z_i^{(t_{ik})}$'s at time T, and where $\hat{\mu}_i$ and $\hat{\sigma}_i^2$ are the empirical means and variances computed by the Rao-Blackwellization (based on the whole chain),

$$\hat{\mu}_i = \frac{1}{T} \sum_{t=1}^{T} \mu_i^{(t)}, \quad \hat{\sigma}_i^2 = \frac{1}{T} \sum_{t=1}^{T} \mu_i^{(t)} \left(1 - \mu_i^{(t)} \right) .$$

We apply this method for the *bIL67* bacteriophage with the single run of 10,000 iterations already used in §6.3.1. To reduce the correlation between the $z_i^{(t)}$'s, we only retain the subsample corresponding to a batch size of 50 with respect to the sequence position i. The corresponding subsample contains 443 points (instead of 22,195). Figure 6.11 illustrates the evolution of the sample (6.2) for several values of T, with the p-value of the Kolmogorov-Smirnov test and the normality plots of Ghosh (1996) (see Chapter 8).

Those results clearly show that normality is not achieved even after $10,000$ iterations. This seems to be paradoxical given the high stability of the allocation states established in the previous section (see the allocation map in Figure 6.7). In fact, the stability in the state allocation implies that the estimated variances $\hat{\sigma}_i^2$ are very close to 0 and those small values are not compensated by a large number N_i of subsampling times (recall that the difference between two subsampling times is distributed as a Poisson variable).This leads to very large values in (6.2) and thus to very large variance for the sample. In our setup, increasing the number of iterations could overcome the drawback induced by the subsampling procedure, as mentioned in §5.6.

FIGURE 6.11. Normality assessments for the average allocation of the states for different values of T, including Ghosh's (1996) normality plots and the Kolmogorov-Smirnov p-value.

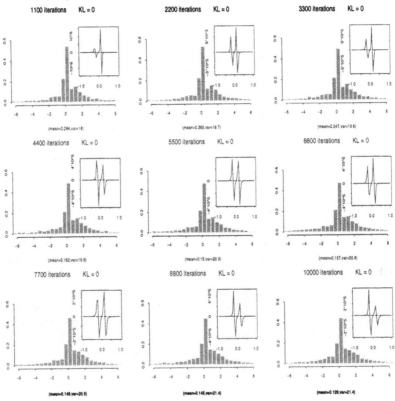

7

Convergence Assessment in Latent Variable Models: Application to the Longitudinal Modelling of a Marker of HIV Progression

Chantal Guihenneuc-Jouyaux
Sylvia Richardson
Virginie Lasserre

7.1 Introduction

Infection[1] with Human Immunodeficiency Virus type-1 (HIV-1), the virus that leads to AIDS, is associated with a decline in CD4 cell count, a type of white blood cell involved in the immune system. In order to monitor the health status and disease progression of HIV infected patients, CD4 counts have thus been frequently used as a marker. In particular, Markov process models of the natural history of HIV play an important part in AIDS modelling (Longini *et al.*, 1991, Freydman, 1992, Longini, Clark and Karon, 1993, Gentleman *et al.*, 1994, Satten and Longini, 1996).

This modelling allows to describe the course of HIV progression in terms of transitions between a certain number of states, which loosely represent various stages of evolution of HIV infection before passage to full blown AIDS. The parameters of the longitudinal Markov model which enable the computation of absorption times from each state to AIDS are the transition rates between the various states. The transition rates can be further cross-classified to evaluate treatment and/or other covariable effects on the progression of the disease.

The classification of a patient in a state is often based on the discretization of values of continuous markers (e.g. CD4 cell count) which are subject to great variability, due mainly to short-term fluctuations of the marker within-subject and to measurement error. The consequences of this vari-

---

[1]We particularly want to thank I.M. Longini and G.A. Satten for their collaboration on this work and C. Monfort for her technical assistance. This work received financial support from the ANRS, contract 096003.

ability are that the observed trajectories of the marker values give a noisy representation of the "true" underlying evolution and consequently, estimating the transition rates, based on raw discretization and not taking into account the short time scale noise, is incorrect. We propose a Bayesian hierarchical model which integrates both a Markov process model and within-individual variability. At a first level, a disease process is introduced as a Markov model on "true" unobserved states corresponding to the disease stages. At a second level, the measurement process linking the true states and the marker values is defined. The quantities of interest (transition rates and measurement parameters) are then estimated with the help of MCMC methods.

A Bayesian hierarchical disease model allowing for misclassification of discrete disease markers has been proposed by Kirby and Spiegelhalter (1994), and we have followed a similar approach. Using a likelihood based estimation method, Satten and Longini (1996), have also considered Markov processes with measurement error with application to modelling marker progression.

Investigation of convergence in this highly dimensional problem with a large number of latent states is challenging. We will illustrate some of the convergence diagnostics introduced in the previous chapters. Since our model involves discrete latent variables, the asymptotic normality convergence diagnostic of Robert, Rydén and Titterington (1998), is particularly appropriate and was found useful in our context.

7.2   Hierarchical Model

The hierarchical model can be conveniently represented with the help of a Directed Acyclic Graph (DAG) linking the unobserved disease states S_{ij} to the observed marker values X_{ij} (CD4 cell counts), where throughout i indexes the individual and j the follow-up point. Different numbers n_i of follow-up points per individual are allowed. Parameters of the disease process are denoted by Λ and δ, those of the measurement process by μ and σ^2. As seen on the graph of Figure 7.1, we have made, in addition to the conditional independence implied by the Markov structure, the following conditional independence assumptions: $[X_{ij} \mid S_{ij}, \mu, \sigma^2]$ is independent of $\{S_{il}, l \neq j\}$ and $\{X_{il}, l \neq j\}$.

7.2.1   Longitudinal disease process

We assume an underlying time homogeneous Markov process with 6 transient states denoted 1 to 6, corresponding to stages of disease progression and based on the CD4 cell count, and a 7th absorbing state corresponding to AIDS and thus recorded without error on the basis of clinical symp-

FIGURE 7.1. Directed Acyclic Graph of the hierarchical model.

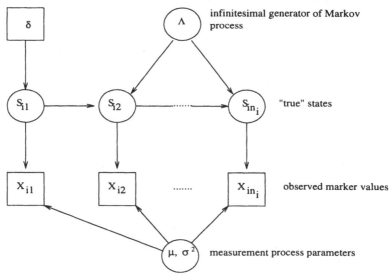

toms. The disease model and associated transition rates are represented in Figure 7.2.

FIGURE 7.2. Markov model of disease progression.

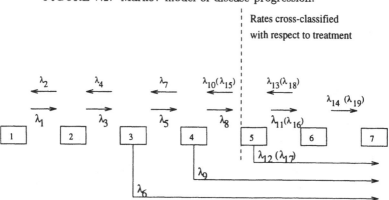

As can be seen, the model allows back flows and direct transitions to AIDS (state 7) from states 3, 4 and 5. Direct transitions were introduced in keeping with clinical knowledge of the infection evolution where sudden accelerated deteriorations of the immune system are observed for some patients. Back flows between adjacent states are allowed as a way of modelling potential immune system improvement after therapy.

As will be detailed when the data set is presented, there was a possibility of therapy for the patients in the most advanced stages. Thus from state 5 we have cross-classified the transition rates, to correspond respectively to treated or untreated follow-up points for the patients. The resulting model has thus 19 transition rates denoted by λ_1 to λ_{19}.

Prior distributions for the transition rates, f_1, are simply taken to be uniform on the interval $[0, 0.25]$, the interval upper bound been chosen large enough to be non informative. A weakly informative first state discrete distribution δ was assumed, since we condition our analysis on the first state.

7.2.2   Model of marker variability

We are concerned here with modelling short time scale fluctuations of CD4 counts, due to inherent within-person variability as well as laboratory measurement errors. An assumption justified by many empirical observations, is to suppose that the variance of the CD4 cell count is better stabilised on a logarithmic scale. Precisely, we suppose that $\log(CD4)$ given the true state (not observed) is Gaussian with unknown mean μ and variance σ^2. In a first approach, the values of $\exp(\mu)'s$ for the 6 states could be taken as center points of intervals of CD4 cell count. Corresponding to clinical practices, the intervals classically considered in the literature are: ≥ 900, $[700, 900[$, $[500, 700[$, $[350, 500[$, $[200, 350[$, $[0, 200[$. In order to relax the model, we chose to consider the means (except the first one which is fixed at $\mu_1 = \log(1100)$) as unknown, but imposing some separation. Precisely, $\exp(\mu_2)$ to $\exp(\mu_6)$ are generated from order statistics on $[100, 1100]$ with a mean spacing equal to 200 CD4.

We considered four different variances

$$\sigma^2 = (\sigma_1^2, \sigma_2^2, \sigma_T^2, \sigma_{NT}^2)$$

for the lognormal distributions to account for higher variability of marker as disease progresses and/or possible influence of treatment effect; σ_1^2 for the first state because it corresponds to an open-ended interval of CD4, σ_2^2 for states 2, 3 and 4 where no treatment is given, and σ_T^2 (respectively σ_{NT}^2) for states 5 and 6 according to whether the patient was treated or not. An assumption of exchangeability for the four variances in σ^2 is made, each one coming from a weakly informative inverse-gamma distribution, f_2, with fixed parameters.

7.2.3   Implementation

The joint posterior distribution of all the parameters was simulated by Gibbs sampling. We list below the full conditional distributions which can be derived from the model assumptions above.

Let us denote by $(\lambda_p, p = 1, ..., 19)$ the 19 transition rates from the model, Λ the matrix of the infinitesimal generator of the Markov process, S_{ij} the true state of patient i at the follow-up time j, n_i the number of follow-up times of patient i, and dt_{ij} the length of the time interval between the $(j-1)th$ and the jth follow-up times of the patient i.

The full conditional distributions in the Gibbs sampler are

$$[\lambda_p \mid \cdot] \quad \propto \quad f_1(\lambda_p) \quad \prod_{i=1}^{n}\prod_{j=1}^{n_i}[S_{ij} \mid S_{ij-1}, dt_{ij}, \Lambda]$$

$$\propto \quad f_1(\lambda_p) \quad \prod_{i=1}^{n}\delta(S_{i1}) \prod_{j=2}^{n_i}\{\exp(\Lambda dt_{ij})\}_{k_1,k_2}$$

where $k_1 = S_{ij-1}$ and $k_2 = S_{ij}$,

$$[S_{ij} \mid \cdot] \propto [S_{ij} \mid S_{ij-1}, dt_{ij}, \Lambda][S_{ij+1} \mid S_{ij}, dt_{ij+1}, \Lambda][\log(X_{ij}) \mid S_{ij}, \sigma^2, \mu]$$

$$[\sigma^{-2} \mid \cdot] \propto f_2(\sigma^{-2}) \quad \prod_{i=1}^{n}\prod_{j=1}^{n_i}[\log(X_{ij}) \mid S_{ij}, \sigma^2, \mu] \tag{7.1}$$

$$[m_k \mid \cdot] \propto [m_k \mid m_{k'\neq k}] \quad \prod_{i=1}^{n}\prod_{j=1}^{n_i}[X_{ij} \mid S_{ij}, \sigma^2, m] \,,$$

where $m_k = \exp(\mu_k)$.

To sample from the non standard full conditional distributions for the λ's and the $\exp(\mu)$'s, we introduced a Metropolis step. The parameter set thus includes all the transition rates, the five means and the four variances. Besides these parameters, the unobserved latent states of the disease process for the 3833 follow-up times are simulated.

7.3 Analysis of the San Francisco Men's Health Study

The formulation of the hierarchical model as applied to the San Francisco Men's Health Study Cohort was part of a collaborative project with Professor I.M. Longini (Emory University, Atlanta).

7.3.1 Data description

The CD4 data on HIV patients of the cohort of San Francisco (the San Francisco Men's Health Study Cohort) has been analysed. This data set consists of 430 male patients monitored approximately every 6 months from mid 1984 through September 1992 and contains 3833 follow-up times with recorded CD4 count, with an average of 8 to 9 time points per patient. The size of this cohort as well as the length of follow-up allow a good characterization of the evolution. Moreover, a subgroup of patients among those having a CD4 cell count lower than 350, received a treatment. This will enable us to test the potential effect of the treatment as it was administered in this cohort (this was not a clinical trial). At each follow-up time, we know if the patient received AZT and/or Pentamidine. We study the treatment effect under an assumption of persistence, i.e. a patient is considered treated at date t if he received a treatment at t or before. Alternative assumptions could be considered.

7.3.2   Results

Table 7.1 presents the estimations of the transition rates with their interval of posterior credibility at 95%. It corresponds to the last 9000 iterations of the MCMC algorithm after a burn-in of 1000 iterations.

A first comment is that the backflows are not negligible, showing the relevance of introducing such transitions in the Markov model. Concerning the treatment effect, we notice that the transition rate λ_{19} for the treated patients from state 6 to state 7 is somewhat smaller than the corresponding one without treatment λ_{14}, but that it is the inverse phenomenon from state 5. Recalling that the treatment was not blindly administered, we venture the explanation that the patients who received treatment early showed more acute clinical signs than the others, and hence being treated at this early stage has operated as a selection phenomenon of a subgroup of more fragile patients, a sort of frailty effect. Moreover, the estimations of the five means μ_2 to μ_6 of the lognormal distribution presented in Table 7.2 are always smaller than the center of the log interval classically used (see §7.2.2). This remark confirms that the last states contain particular selected patients with very small CD4 cell count.

In order to measure the potential treatment effect on the progression of the disease, a new parameter $\theta_{6\to7}$ can be introduced namely the ratio of the transition rate for treated versus untreated. For example, concerning transition from state 6 to AIDS, we calculate at iteration t of the MCMC algorithm,

$$\theta_{6\to7}^{(t)} = \frac{\lambda_{19}^{(t)}[treatment]}{\lambda_{14}^{(t)}[without\ \ treatment]}.$$

Thus, if the treatment is effective, the ratio $\theta_{6\to7}$ should be lower than 1. Figure 7.3 gives the posterior distribution of $\theta_{6\to7}$ based on the last 9000 iterations. The mean of $\theta_{6\to7}$ is 0.83 and its posterior credibility interval $[0.41, 1.11]$, giving some indication of a treatment effect for slowing late passage to AIDS.

An American study (Satten and Longini, 1996), made on the same data with a different hierarchical model and with maximum likelihood estimations (profile likelihood), showed a stronger treatment effect with θ equal to 0.44 with confidence interval $[0.31, 0.61]$. As to be expected, the interval of variability obtained with our Bayesian model is larger than that ob-

TABLE 7.1. Estimation of the transition rates (in month^{-1}).

		Without treatment		With treatment
$1 \rightarrow 2$	λ_1	0.038		
		[0.027, 0.052]		
$2 \rightarrow 3$	λ_3	0.032		
		[0.026, 0.039]		
$3 \rightarrow 4$	λ_5	0.047		
		[0.038, 0.056]		
$4 \rightarrow 5$	λ_8	0.041		
		[0.034, 0.049]		
$5 \rightarrow 6$	λ_{11}	0.036	λ_{16}	0.054
		[0.023, 0.050]		[0.032, 0.083]
$6 \rightarrow 7$	λ_{14}	0.172	λ_{19}	0.112
		[0.111, 0.240]		[0.080, 0.154]
$3 \rightarrow 7$	λ_6	0.003		
		[0.001, 0.005]		
$4 \rightarrow 7$	λ_9	0.002		
		[0.0002, 0.006]		
$5 \rightarrow 7$	λ_{12}	0.013	λ_{17}	0.012
		[0.002, 0.025]		[0.001, 0.028]
$2 \rightarrow 1$	λ_2	0.003		
		[0.001, 0.006]		
$3 \rightarrow 2$	λ_4	0.005		
		[0.002, 0.009]		
$4 \rightarrow 3$	λ_7	0.013		
		[0.007, 0.021]		
$5 \rightarrow 4$	λ_{10}	0.013	λ_{15}	0.004
		[0.006, 0.023]		[0.0003, 0.013]
$6 \rightarrow 5$	λ_{13}	0.014	λ_{18}	0.008
		[0.0004, 0.047]		[0.0003, 0.033]

tained through profile likelihood. Indeed, by using a joint model of disease and marker variability, the fluctuation of measurement process parameters are fully propagated on the estimations of the underlying transition rates. Here, this leads to less positive conclusions on treatment effect than using a classical approach.

From the estimation of the transition rates, it is possible to calculate absorption times to the AIDS state (state 7) starting from a given state. Table 7.3 gives these times expressed in years.

It seems surprising that the absorption times to AIDS, except from state

TABLE 7.2. Estimations of the five means and their 95% credibility interval.

	Estimation of the mean	Center of classical log interval
μ_2	6.59 $[6.56, 6.61]$	6.67
μ_3	6.25 $[6.21, 6.29]$	6.38
μ_4	5.86 $[5.81, 5.91]$	6.04
μ_5	5.28 $[5.20, 5.37]$	5.58
μ_6	4.02 $[3.92, 4.16]$	4.61

FIGURE 7.3. Posterior distribution of $\theta_{6 \to 7}$.

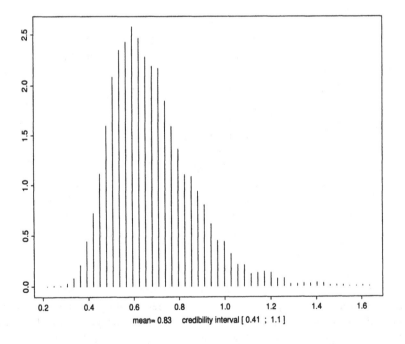

mean= 0.83   credibility interval [0.41 ; 1.1]

6, have a tendency to be smaller for the treated patients than the others. This is due to the paradoxical faster transition from state 5 to 6 for the treated patients and the potential frailty effect which we discussed earlier.

TABLE 7.3. Absorption times (in year) to AIDS.

Stage	Without treatment	With treatment
1	11.9	11.3
	[10.9, 13.1]	[10.2, 12.6]
2	9.7	9.1
	[8.8, 10.6]	[8.2, 10.1]
3	6.8	6.2
	[6.2, 7.6]	[5.5, 7.1]
4	5.1	4.5
	[4.5, 5.9]	[3.8, 5.3]
5	2.8	2.1
	[2.3, 3.5]	[1.7, 2.8]
6	0.7	0.9
	[0.4, 1.1]	[0.6, 1.2]

7.4   Convergence assessment

The full conditional distributions of the transition rates in the Gibbs sampling involve computing the exponential of the infinitesimal generator matrix Λ, i.e.

$$\prod_{i=1}^{n} \prod_{j=2}^{n_i} \exp\{\Lambda dt_{ij}\},$$

where n_i is the number of follow-ups for patient i and dt_{ij} is the time interval between follow-ups $j-1$ and j. The exponential of the matrix (Λdt_{ij}) was computed using a diagonalisation routine. To sample from this non standard distribution required an additional Metropolis– Hastings step which we implemented with a random walk proposal. The size of the matrix Λ (19x19) and the simulation of a large number of unobserved states lead to very long running times which prevent from using parallel runs for assessing convergence. The diagnostics studied below are thus applied to a single run of 10,000 iterations.

Firstly, the different diagnostics of CODA can be applied to the output chain. These diagnostics have been detailed in Chapter 6. Figures 7.4 to 7.6 give examples of output from CODA. Raftery and Lewis' (1992a) evaluation suggests between 10 and 200 iterations for the warm-up time and between 4000 and 240, 000 iterations for convergence time. As in most cases, the proposed warm-up time seems to be overly optimistic and some of the run lengths indicated somewhat conservative. To illustrate different patterns of convergence, we chose to display a set of typical parameters for which Raftery and Lewis' (1992a) evaluation indicated contrastingly large or moderate number of iterations: for the transition rates, λ_1, λ_5 and

FIGURE 7.4. Plots of the simulation output for the parameters of interest in the HIV model, based on 10,000 iterations, obtained by CODA. (The parameters are, from top to bottom, λ_1, λ_5, λ_{10}, λ_3, λ_4, λ_{12}, σ_T^2, σ_{NT}^2 and σ_2^2.)

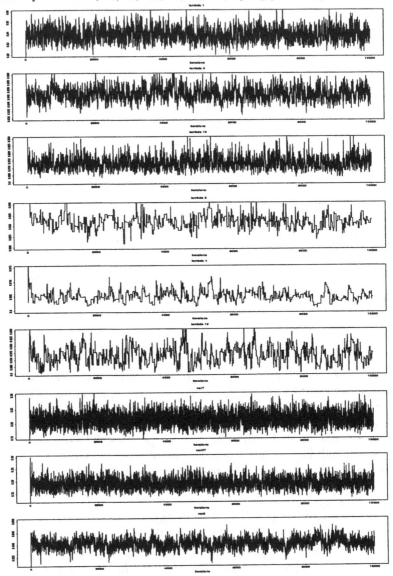

λ_{10} up to 15,000 iterations while for λ_3, λ_4 and λ_{12} at least 100,000 iterations. Similarly, for the variance parameters, σ_T^2 and σ_{NT}^2 required less than 10,000 iterations against σ_2^2 for which about 25,000 iterations are needed. Trace plots for these transition rates and variances are shown in

Figure 7.4. We see no indication of poor mixing performance of the sampler even though slower mixing occurs for the same transition rates (λ_3, λ_4 and λ_{12}) as detected by Raftery and Lewis' (1992a) diagnostic. Note that for reasons of practicality, we did not tune the proposal separately for each transition rate. Figure 7.4 indicates that it could be necessary for these latter three rates.

FIGURE 7.5. Geweke's (1992) diagnostic plot for the parameters of interest in the HIV model, based on 10,000 iterations, obtained by CODA. (The parameters are denoted by varT for σ_T^2, varNT for σ_{NT}^2 and var2 for σ_2^2.)

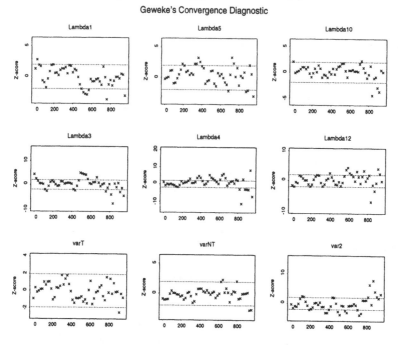

Figure 7.5 describes Geweke's (1992) diagnostic based on 10,000 iterations thinned by a factor 10 for the same set of parameters. There is reasonable stability but we note that the scale for the Z-scores is sensibly larger for the same three rates as well as σ_2^2, with quite a few points outside the confidence region, in agreement with the previous remarks. Figure 7.6 gives the autocorrelograms for these parameters based on 10,000 iterations (thinned by a factor 10). These autocorrelation plots give indications somewhat similar to the previous diagnostics, with higher autocorrelations for λ_3, λ_4, λ_{12} and σ_2^2 than the others (precisely, the autocorrelations are significant up to lag 10 and only up to lag 3 for the others). Heidelberger and Welch's (1983) diagnostic does not detect convergence problems for the 19 parameters of interest after 10,000 iterations (results not shown).

FIGURE 7.6. Autocorrelograms for selected parameters based on 10,000 iterations (thinned by a factor 10), obtained by CODA. (The parameters are denoted by varT for σ_T^2, varNT for σ_{NT}^2 and var2 for σ_2^2.)

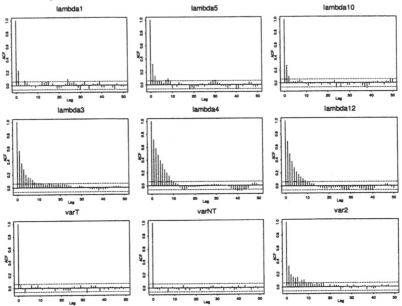

These diagnostics thus have not detected a serious lack of convergence but they give rather weak information on run lengths, except for Raftery and Lewis' (1992a) evaluation which may indicate extreme numbers of iterations. Overall, except for Heidelberger and Welch's (1983) diagnostic, they were concordant in highlighting a subgroup of parameters for which convergence was slower.

Another approach is to use the asymptotic normality diagnostic of Robert, Rydén and Titterington (1998) presented in §5.6, based on the latent variables in the model, that is the unobserved states, as in the DNA application of Chapter 6. This has the additional interest of producing a global control which is easier to interpret than the separate monitoring of each parameter. These states are simulated at each follow-up time conditionally on their neighbours, due to the Markov structure of the disease process. If S_{ij}^t is the unobserved state of patient i at the jth follow-up time $(j \geq 1)$ and at iteration t of the MCMC algorithm, then S_{ij}^t has a discrete distribution, conditional on S_{ij-1}^t, S_{ij+1}^t and the current values of the parameters (see (7.1)), on state space $\{1, 2, 3, 4, 5, 6\}$. For the asymptotic normality diagnostic, the MCMC chains of S_{ij}^t's are subsampled at random times (for each couple (i, j) taking in total 3833 values), as discussed in §5.6. (More precisely, the difference between these times is generated by a Poisson distribution.) At the end of the MCMC run (T iterations), a sample of 3833 normalized sums, $\tilde{S}_T$ (see (5.26)), is computed with mean and variance estimated from the complete run of the MCMC algorithm.

As noted in Robert *et al.* (1998), the random subsampling does not eliminate the correlation between the S_{ij}^t's induced by the longitudinal structure. Therefore if we simultaneously consider the 3833 values, asymptotic normality would be perturbed. We thus only consider a subset corresponding in this case to the last follow-up time for each patient (containing 430 points). Other choices give similar results.

In order to illustrate how normality is improved as T increases, Figure 7.7 presents histograms of $\tilde{S}_T$ and normality plots via the T_3-function[2] of Ghosh (1996) (see Chapter 8) for T between 1000 and 9000. The asymptotic normality becomes acceptable for T greater than 6000 iterations, with a Kolmogorov–Smirnov p-value equal to 0.42 for $T = 6000$ and to 0.75 for $T = 10,0000$. Nevertheless, even though the plot of the T_3-function is gradually modified so that it stays within the confidence limits when $T = 10,000$ iterations, it still does not compare with a straight line. This control shows that more than $10,000$ iterations are thus necessary for achieving approximate normality.

---

[2] The graphical normality assessment of Ghosh (1996) is based on the properties of the third derivative of the logarithm of the empirical moment generating function, called T_3-function, in the normal case

FIGURE 7.7.　Convergence control by asymptotic normality assessment. The histograms of the samples of $\tilde{S}_T$'s are represented, along with the normality T_3-function plots of Ghosh (1996), including 95% *(dashes)* and 99% *(dots)* confidence regions.

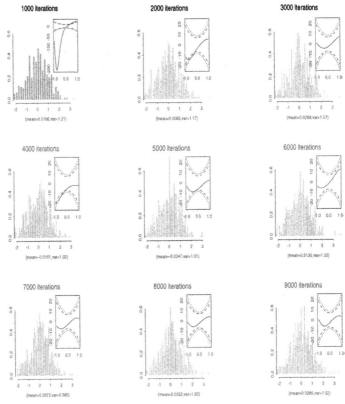

8
Estimation of Exponential Mixtures

Marie-Anne Gruet
Anne Philippe
Christian P. Robert

8.1 Exponential mixtures

8.1.1 Motivations

Exponential mixtures are distributions of the form

$$\sum_{i=0}^{k} p_i \, \mathcal{E}xp(\lambda_i), \tag{8.1}$$

with $p_0 + \ldots + p_k = 1$ and $\lambda_i > 0$ $(0 \leq i \leq k)$. Considering the huge literature on normal mixtures (see §3.4), the treatment of exponential mixtures is rather limited. A possible reason, as illustrated in this chapter, is that the components of (8.1) are much more difficult to distinguish than in the normal case of §3.4. Exponential mixtures with a small number of components are nonetheless used in the modeling of phenomena with positive output and long asymmetric tails, mainly in survival and duration setups, like the applications mentioned in Titterington, Smith and Makov (1985, p.17-21). We also illustrate this modeling in the case of hospitalization durations for which a two or three component exponential mixture is appropriate.

As in the normal mixture example of Chapter 3, exponential mixtures have very interesting features from an MCMC point of view.[1] We first show in §8.1.2 how a reparameterization of the model (8.1) can provide a noninformative prior with well-defined posterior. We then discuss the MCMC implementation associated with this posterior, with strong emphasis on the convergence diagnostics developed in the previous chapters or peculiar to mixture models.

The modeling and diagnostics are tested on both a simulated sample of 250 observations, generated from the exponential mixture

$$0.5 \, \mathcal{E}xp(1) + 0.38 \, \mathcal{E}xp(0.66) + 0.12 \, \mathcal{E}xp(0.33) \tag{8.2}$$

---

[1] See Mengersen, Robert and Guihenneuc–Jouyaux (1998) for a comparative study of diagnostic methods in the setup of mixtures.

and on a real dataset of 2292 observations, which corresponds to hospitalization durations (in days) in the geriatric section of an hospital in Grenoble (south of France), from 1994 to 1995. The data is heterogeneous since it covers regular geriatric care (short stays), convalescence stays and intensive care (longer), although this administrative division does not necessarily correspond to homogeneous subgroups.[2]

8.1.2 Reparameterization of an exponential mixture

As in the study of a normal mixture in §3.4, the parameterization of a mixture does matter, both in terms of selection of a prior distribution and of implementation of an MCMC algorithm for the approximation of the corresponding Bayes estimates. In the case of an exponential mixture (8.1), a possible reparameterization is a scale representation of the λ_i's, that is $(i = 1, \ldots, k)$

$$\lambda_i = \lambda_0 \tau_1 \ldots \tau_i,$$

with a similar expansion on the p_i's, i.e. $p_0 = q_0$ and $(i = 1, \ldots, k-1)$

$$p_i = (1 - q_0) \ldots (1 - q_{i-1}) q_i.$$

Already used for normal mixtures, this reparameterization is oriented towards a parsimonious representation of mixtures, in the sense that a k component mixture appears as a modification of a $(k-1)$ component mixture, since the last component of the $(k-1)$ component mixture,

$$\mathcal{E}xp(\lambda_0 \tau_1 \ldots \tau_{k-2}),$$

is replaced with a two component exponential mixture,

$$q_{k-2} \mathcal{E}xp(\lambda_0 \tau_1 \ldots \tau_{k-2}) + (1 - q_{k-2}) \mathcal{E}xp(\lambda_0 \tau_1 \ldots \tau_{k-1}).$$

This representation of exponential mixtures can thus be interpreted as ranking the components by order of importance or, rather, of breadth, in the sense that the first component corresponds to the global parameter of the model, λ_0, while the following components correspond to more and more local departures from the original exponential distribution $\mathcal{E}xp(\lambda_0)$. This representation is particularly interesting in the case k is unknown, as mentioned in §8.1.3, since it prevents strong increases in the number of components.

The parameterization (τ_i, q_i) also facilitates the derivation of a noninformative prior distribution, when compared with the original parameterization (λ_i, p_i). Indeed, it is well-known in mixture setups that improper

---

[2] Thanks to Gilles Celeux for introducing us to this problem and also for providing the hospitalization dataset!

priors of the form

$$\prod_i \pi(\lambda_i),$$

cannot be used, the basic reason being that there is always a positive probability that no observation is allocated to one of the components. On the opposite, if some dependence structure is introduced among the λ_i's, the overall prior can still be improper while the corresponding posterior distribution remains proper (see also Robert and Titterington, 1998). For instance, the choice

$$\pi(\lambda_0, \tau_1, \ldots, \tau_k, q_0, \ldots, q_{k-1}) \propto \frac{1}{\lambda_0} \mathbb{I}_{\tau_1 \leq 1} \ldots \mathbb{I}_{\tau_k \leq 1} \mathbb{I}_{q_0 \leq 1} \ldots \mathbb{I}_{q_{k-1} \leq 1} \quad (8.3)$$

leads to a well-defined posterior distribution. The uniform priors on the τ_i's are justified through the identifiability constraint

$$\lambda_0 \geq \lambda_1 \geq \ldots \geq \lambda_k. \quad (8.4)$$

Lemma 8.1.1 *The posterior distribution associated with the improper prior distribution (8.3) is a proper distribution for every possible sample.*

Proof. That the posterior distribution associated with the improper prior distribution (8.3) and the exponential mixture (8.1) is finite for any sample size can be shown by integrating $\pi(\lambda_0, \tau_1, \ldots, \tau_k, q_0, \ldots, q_{k-1}|x, z)$ for every configuration $z = (z_1, \ldots, z_n)$. If i_0 (i_1) is the smallest (largest) component number such that $n_i > 0$,

$$\int \pi(\lambda \ldots, q_{k-1}|x, z)d\lambda = \int \frac{1}{\lambda_0^2} \frac{1}{\lambda_1} \ldots \frac{1}{\lambda_{k-2}} \lambda_{i_0}^{n_{i_0}} e^{-s_{i_0}\lambda_{i_0}} \ldots \lambda_{i_1}^{n_{i_1}} e^{-s_{i_1}\lambda_{i_1}} d\lambda$$

$$= \int \frac{1}{\lambda_{i_0}^2} \ldots \frac{1}{\lambda_{i_1-1}} \lambda_{i_0}^{n_{i_0}} e^{-s_{i_0}\lambda_{i_0}} \ldots \lambda_{i_1}^{n_{i_1}} e^{-s_{i_1}\lambda_{i_1}} d\lambda_{i_0} \ldots d\lambda_{i_1}$$

$$\leq K \int \frac{1}{\lambda_{i_0}^2} \lambda_{i_0}^{n_{i_0}} e^{-s_{i_0}\lambda_{i_0}} \ldots \lambda_{i_1-1}^{n_{i_1-1}-1} e^{-s_{i_1-1}\lambda_{i_1-1}} \lambda_{i_1-1}^{3/2} d\lambda_{i_0} \ldots d\lambda_{i_1-1}$$

by Cauchy-Schwarz inequality, where K is a finite normalizing constant. Therefore,

$$\int \pi(\lambda \ldots, q_{k-1}|x, z)d\lambda \leq K' \int \frac{1}{\lambda_{i_0}^2} \lambda_{i_0}^{n_{i_0}} e^{-s_{i_0}\lambda_{i_0}} \lambda_{i_0}^{3/2} d\lambda_{i_0}$$

$$= K' \int \lambda_{i_0}^{n_{i_0}-1/2} e^{-s_{i_0}\lambda_{i_0}} d\lambda_{i_0} < \infty.$$

$\Box\Box$

8.1.3   Unknown number of components

An extension of the standard mixture inferential problem deals with the estimation of the x number of components, $k+1$. This problem has been addressed by Richardson and Green (1997) in the case of normal mixtures and by Gruet, Philippe and Robert (1998) for exponential mixtures, through the technique of reversible jump MCMC, recalled in §1.2. The additional steps, when compared with the Gibbs sampler $[A_{20}]$ presented below, are the split and merge moves, which change the number of components in the mixture (8.1) to $k+2$ or k. The parameterization in (τ_i, q_i) leads to straightforward solutions. Indeed, in the move from component i_0 to components i_0 and $i_0 + 1$, the new parameters are naturally chosen as

$$\tau'_{i_0} = u_1 + \tau_{i_0}(1 - u_1), \qquad \tau'_{i_0+1} = \tau_{i_0}/(u_1 + \tau_{i_0}(1 - u_1)),$$

and

$$q'_{i_0} = q_{i_0} u_2, \qquad q'_{i_0+1} = \frac{q_{i_0}(1 - u_2)}{1 - q'_{i_0}},$$

where u_1 and u_2 are uniform on $(0, 1)$ when $i_0 > 0$, and u_1 is uniform on $(.5, 1)$ for $i_0 = 0$, with

$$\tau'_0 = \tau_0/u_1, \qquad \tau'_1 = u_1.$$

8.1.4   MCMC implementation

As in the case of the normal mixture of §3.4, a Gibbs sampler associated with the prior (8.3) is straightforward to implement. The algorithm is a *Data Augmentation* type scheme which is, again, based on the completed model, associated with the (z_i, x_i)'s, where z_i is the component indicator such that $x_i|z_i \sim \mathcal{E}xp(\lambda_{z_i})$. The two steps in the algorithm are then

1. Complete the missing data by $(i = 0, \ldots, k,\ j = 1, \ldots, n)$

$$P(Z_j = i) \propto p_i \lambda_i \exp(-\lambda_i x_j)$$

2. Generate $[A_{20}]$

(a) $\lambda_0 \sim \mathcal{G}a(n, \sum_{j=0}^{k} \tau_1 \ldots \tau_j n_j \bar{x}_j)$;

(b) $\tau_i \sim \mathcal{G}a(n_i + \ldots + n_k + 1, \sum_{j=i}^{k} \frac{\lambda_j}{\tau_i} n_j \bar{x}_j) \mathbf{I}_{\tau_i < 1}$; $i = 1, \ldots, k$

(c) $q_i \sim \mathcal{B}e(n_i + 1, n_{i+1} + \ldots + n_k)$; $i = 0, \ldots, k - 1$

where n_j denotes the number of observations allocated to the component j and $\bar{x}_j$ is the mean of these observations (with the usual convention that $n_j \bar{x}_j = 0$ when $n_j = 0$). As pointed out by Robert and Titterington (1998), in the normal case, a Gibbs sampler is not available for the parameterization of Robert and Mengersen (1998), but only for the original parameterization. This is not the case for exponential mixtures.

Note that the subchain (z_t) depends on the previous value of the parameters. Therefore, the subchain (z_t) is not a Markov chain and the Duality Principle of §1.5 does not apply. It is then impossible to take advantage of the finite state space chain (z_t) in the usual way.

In $[A_{20}]$, step 2.(b) involves truncated gamma distributions which can be generated using the optimal accept-reject algorithm of Philippe (1997c) or Damien and Walker's (1996) device of auxiliary variables, representing the density

$$\pi(\tau) \propto \tau^{\alpha-1} e^{-\beta\tau} \, \mathbb{I}_{\tau<1}$$

as the marginal of

$$\pi(\tau,\omega) \propto \tau^{\alpha-1} \, \mathbb{I}_{\tau<1} \, \mathbb{I}_{\omega<\exp(-\beta\tau)} \, .$$

This second approach induces an additional stage in the above Gibbs sampler, as follows:

(b.1) **Generate**

$$\omega^{(t)} \sim \mathcal{U}([0, e^{-\beta\tau}]);$$

(b.2) **Generate**

$$\tau^{(t)} \sim \tau^{\alpha-1} \, \mathbb{I}_{-\log(\omega)/\beta<\tau<1} \, .$$

8.2 Convergence evaluation

8.2.1 Graphical assessment

As already mentioned for normal mixtures, the problem of diagnosing convergence is particularly challenging in mixture settings and specific diagnostics must be constructed accordingly, since empirical evaluations are often found to be over-optimistic. If we add the missing data vector to the parameter vector, the dimension of the state space $n + 2k + 2$ leads to large dimension Markov chains, for which intuition is most usually at a loss and which may mix very slowly, thus requiring large numbers of iterations.

The standard graphical tools, which evaluate whether or not convergence has occurred, can be invoked in this setup, namely, after a warmup run of n iterations,

(a) a comparison of the sample histogram with the estimated density (or a nonparametric alternative).

(b) As suggested by Aitkin (1997), a comparison of the empirical cdf with the estimated cdf, along with a quantitative assessment of the fit (for instance, through a Kolmogorov-Smirnov test), even though this criterion does not evaluate the adequation in the tails.

(c) The plot of the allocations averaged along the iterations,

$$z_j^T = \frac{1}{T} \sum_{t=1}^{T} z_j^{(t)}, \qquad (8.5)$$

against the estimated expected allocation $\mathbb{E}[z_j|x_j, \theta^{(T)}]$, which may often point out different mixing rates in the allocation chain and in the parameter chain. It may also signal the inadequacy of the mixture modeling.

(d) The evolution of the allocations z_i along the iterations, called *allocation map* and already presented in the normal mixture case (see Figure 3.8).[3]

Figure 8.1 contains a spreadsheet corresponding to these different criteria, obtained after a Gibbs sampling run of 50,000 iterations, for the simulated sample of size 250 from the three component exponential mixture (8.2).

The fit of both the density and the cdf are rather satisfactory, as also shown by the high value of the Kolmogorov-Smirnov p-value. (Note that an unconstrained estimation of the mixture does not perform as well, although it provides a useful starting value, as in Robert and Mengersen, 1998.) On the opposite, the allocation map, which gives a graphical representation of the successive allocations of every observation of the sample by allocating a grey level to each component (dark for component 0, grey for component 1 and light grey for component 2), does not show a sign of stabilization of the allocation sequence (for 50,000 iterations), since the horizontal bars in the allocation map indicate that most of the sample is allocated to one component or another, with this component varying along iterations. When considering the empirical averages, Figure 8.2 shows an overall stability in the approximation of the parameter estimates, even though λ_0 exhibits a huge jump after 20,000 iterations. (Note how the sum of the two first weights gets stable more quickly than the first weight, thus exhibiting a kind of compensation phenomenon between the two first components.) The behaviour of the allocation graph thus stresses the weak identifiability structure of the mixtures of exponential distributions, in the sense that

---

[3]This representation could be called the *Griddy Diagnostic*, given the computing time and space requirements it involves!

FIGURE 8.1. A control spreadsheet for the MCMC algorithm associated with the estimation of the exponential mixture, based on a sampler of 250 observations from (8.2). *Upper left:* Histogram of the simulated dataset, fit by the 3 component exponential mixture $0.51\,\mathcal{E}xp(1.24) + 0.26\,\mathcal{E}xp(0.70) + 0.23\,\mathcal{E}xp(0.46)$, and weighted components. *Upper right:* Empirical cdf of the sample and comparison with the estimated cdf. *Lower left:* Comparison of the average allocation (8.5) (full line) and of the estimated expected allocation (8.6) (dashes). *Lower right:* Allocation map of the 250 observations along the 50,000 MCMC iterations. The lower graphs correspond to ordered observations. (*Source:* Gruet *et al.*, 1998.)

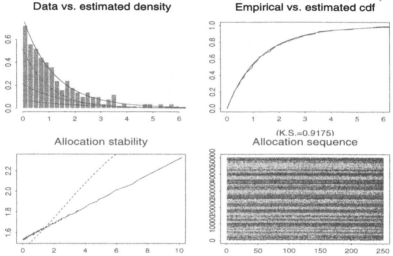

most observations can be indifferently allocated to one component or another, as opposed to normal mixtures where the allocation maps exhibit much more stability (see §3.4 and Robert, 1997, for illustrations).

This feature is also exposed in the control device of Figure 8.1 *[lower left]*, namely the graph of the average allocations versus the expected allocation evaluated on the estimated parameters. Since both quantities should be similar, even though (8.5) also takes into account the posterior variability on the estimated coefficients, an evaluation of convergence can be based on the difference between them, namely the plot of (8.5) against the plot of the expectations

$$\frac{\sum_{i=0}^{k} i\hat{p}_i \hat{\lambda}_i e^{-x_j \hat{\lambda}_i}}{\sum_{i=0}^{k} \hat{p}_i \hat{\lambda}_i e^{-x_j \hat{\lambda}_i}},\qquad (8.6)$$

where the estimates $\hat{\lambda}, \hat{\sigma}, \hat{p}$ are the ergodic averages (or their Rao-Blackwellized alternatives) after T iterations. The two plots are markedly different, with allocations more central than the expected values. Although (8.5) does not converge to (8.6) with T, the difference is much larger than in the normal case (see, e.g., Robert, Rydén and Titterington, 1998) and indicates a wider range for the quantities

$$\frac{\sum_{i=0}^{k} i \, p_i \lambda_i e^{-x_j \lambda_i}}{\sum_{i=0}^{k} p_i \lambda_i e^{-x_j \lambda_i}} \, ,$$

than for (8.5).

FIGURE 8.2. Convergence of the empirical averages of the different parameters of a three component exponential model for a sample of 250 observations simulated from (8.2). The weights are represented through their cumulated sums p_1 and $p_1 + p_2$, and the λ_i's are ranked according to the identifiability constraint. (*Source: Gruet et al.*, 1998.)

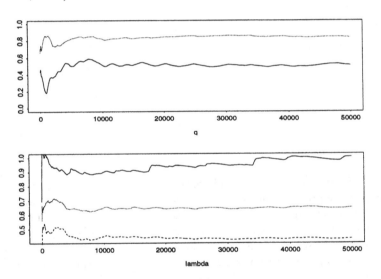

In the case of the hospitalization dataset, Figure 8.3 provides the overall fit diagnostic for a 4 component exponential mixture, which is qualitatively satisfactory despite a null p-value for the Kolmogorov-Smirnov test. (This value may be attributed to the large sample size, rather than to strong departures of the estimate cdf from the observed cdf.) Note also the stable features in the allocation sequence which appear after a warm-up period of about 1000 iterations. The averages in Figure 8.4 are rather stable, if not entirely so. The agreement between (8.5) and (8.6) is much stronger than in the simulated example, but this may be due to the large weight of the first component (see Figure 8.3 *[Upper left]*).

FIGURE 8.3. Control spreadsheet for the hospital stay dataset (same legend as Figure 8.1). (*Source:* Gruet *et al.*, 1998.)

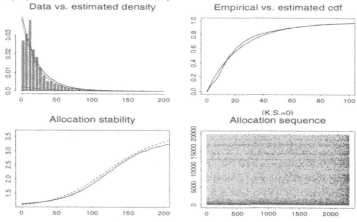

FIGURE 8.4. Convergence of the parameters for the hospital stay dataset (same legend as Figure 8.2). (*Source:* Gruet *et al.*, 1998.)

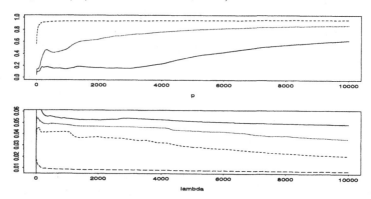

8.2.2 Normality check

As in §5.6, we also consider an assessment of convergence via a normality check on the average allocations $(j = 1, \ldots, n)$.

$$z_j^T = \frac{1}{10T} \sum_{t=1}^{T} z_j^{(10t)}, \qquad (8.7)$$

which are standardized via empirical means and variances computed by Rao-Blackwellization,

$$\hat{\mu}_j = \frac{1}{T} \sum_{t=1}^{T} \frac{\sum_{i=1}^{k} i p_i e^{-\lambda_i x_j}}{\sum_{i=1}^{k} p_i e^{-\lambda_i x_j}}, \qquad \hat{\sigma}_j^2 = \frac{1}{T} \frac{\sum_{i=1}^{k} i^2 p_i e^{-\lambda_i x_j}}{\sum_{i=1}^{k} p_i e^{-\lambda_i x_j}} - \hat{\mu}_j^2.$$

While the $z_j^{(t)}$'s are not (marginally) independent, subsampling with a batch size of 10 in (8.7) brings us closer to independence and also diminishes the dependence between the z_j^T's and the estimators $(\hat{\mu}_j, \hat{\sigma}_j)$.

An exact evaluation can be based on the Central Limit Theorem as shown by Robert *et al.* (1998), following Theorem 5.6.1. A random subsampling of the $z_j^{(t)}$'s with increasingly distant sampling times indeed produces an asymptotically $\mathcal{N}(0, 1)$ sample. A normality check on the standardized and averaged allocations (8.7) thus provides a first approximation (or a limiting case, for $d = 0$ in Theorem 5.6.1) to the method developed in §5.6.

For the simulated sample with 250 observations generated from (8.2), Figure 8.5 shows the evolution of the sample of the

$$\sqrt{10T} \frac{z_j^T - \hat{\mu}_j}{\hat{\sigma}_j}, \qquad j = 1, \dots, n, \tag{8.8}$$

against T, along with the successive normality plots of Ghosh (1996) and the p-value of the Kolmogorov-Smirnov test of normality. The graphical normality assessment of Ghosh (1996) is based on the properties of the third derivative of the logarithm of the empirical moment generating function, called the T_3−*function*, in the normal case. Deviation of the curve of the T_3−function from the horizontal zero line indicates lack of normality. Moreover, a Central Limit Theorem for the T_3−function provides approximate confidence bands such that a departure of normality is spotted at a given significance level if the curve of the T_3−function crosses the corresponding upper or lower bounds anywhere in the interval $[-1, 1]$. The uniform good performance of the sample for this control method seems paradoxical, given the lack of stability of the allocation graphs in Figure 8.1, but this may be a signal of the good mixing properties of the algorithm. In the normal case, where allocation graphs are much more stable (see, e.g., Robert *et al.*, 1998), the control based on the Central Limit Theorem requires a much larger number of iterations.

The results of the normality test for the hospitalization dataset, given in Figure 8.6, are very satisfactory. The fact that the first component takes more than 80% of the whole mass may explain for this higher stability, when compared with the simulated example.

8.2.3 Riemann control variates

The convergence control based on the random Riemann sums (see §3.3) applies in this setup. Note that, for the parameterization in $\zeta = (\lambda, \tau, p) \in$

FIGURE 8.5. Convergence control, for the simulated sample, based on normality assessments for the normalized allocations (8.8) for several values of T:
each graph includes the histogram of the sample, plotted against the normal cdf,
and Ghosh (1996) normality plot, with the corresponding Kolmogorov-Smirnov
p-value. (*Source:* Gruet *et al.*, 1998.)

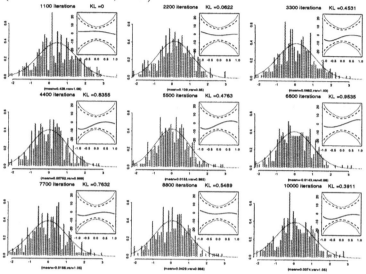

FIGURE 8.6. Normality assessment on the allocations, for the hospitalization
dataset (same legend as Figure 8.5). (*Source:* Gruet *et al.*, 1998.)

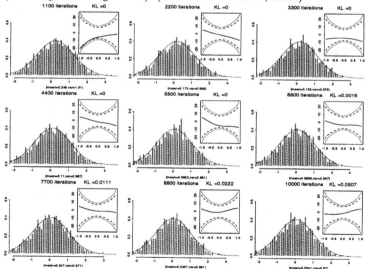

$\mathbb{R}^{2k-1}$, it is not possible to obtain an explicit form of the marginal densities. Thus, the standard Riemann estimator (3.3) cannot be used. However, since the densities of the full conditional distribution are available, the Rao-Blackwellized version of the Riemann sums can be computed. Note that the estimates are built for the parameters $(\lambda_0, \tau_1, \tau_2, q_0, q_1)$, since this is the parameterization which leads to closed form Rao-Blackwellized estimates.

Indeed, the expectation $\mathbb{E}[h(\lambda, \tau, p)]$ can be expressed as

$$\int h(\lambda, \tau, p)\pi(\lambda, \tau, p)\, d\lambda\, d\tau\, dp = \int h(\zeta)\pi(\zeta^\ell|\zeta^{[\ell]})\pi(\zeta^{[\ell]})d\zeta$$

for every fixed $\ell = 1, \ldots, 2k - 1$ and with $\zeta^{[\ell]} = (\zeta^r, r \neq \ell)$. Therefore, for a given function h, we can construct $2k - 1$ different Rao-Blackwellized Riemann estimators, namely the estimators ($\ell = 1, \ldots, 2k - 1$)

$$\delta_\ell^h = T^{-1}\sum_{t=1}^{T-1}(\zeta^\ell_{(t+1)} - \zeta^\ell_{(t)})\left\{\sum_{b=1}^{T}h(\zeta^\ell_{(t)}, \zeta^{[\ell]}_b)\pi(\zeta^\ell_{(t)}, \zeta^{[\ell]}_b)\right\}, \qquad (8.9)$$

where

$$\zeta^\ell_{(1)} \leq \cdots \leq \zeta^\ell_{(t)} \leq \cdots \leq \zeta^\ell_{(T)}$$

denotes the order statistics associated with the ℓ-th component of the MCMC sample $\zeta^{(1)}, \ldots, \zeta^{(T)}$.

By taking $h(\zeta) = \zeta^m$ ($m = 1, \ldots, 2k-1$), we thus get $(2k-1)$ convergent estimators of each parameter of the mixture model which can be used through a simple convergence diagnostic, namely the agreement of the $(2k-1)$ evaluations δ_ℓ^m (in ℓ) as in usual control variate methods based on the common stabilization of different estimates of the same quantity (see §2.2.1 and §3.3.3). Moreover, the control variate control technique, based on the estimation of the constant function $h(\zeta) = 1$ is still available (§3.3.3). This technique allows in particular for an evaluation of the correct coverage of the support of π, since it detects chains which have failed to explore significant regions of the support of π, if the estimate of the marginal distribution in (8.9) does not bias the evaluation.

As shown in Figure 8.7, for the simulated sample, the various parameters lead to different convergence times in the sense that the control variate estimates δ_ℓ^1 reach the 1% or 5% error bound for different numbers of iterations, but they all converge to 1. The Rao-Blackwellized Riemann estimates of the different parameters, corresponding to different choices of ℓ in (8.9), all lead to the same value in a small number of iterations.

For the hospital stay dataset, Figure 8.8 illustrate the convergence of the Rao-Blackwellized Riemann control variates, which guarantee that the whole supports of the marginal densities of λ and q_i's have been visited. Note that, in both cases, a closed form of the marginal densities of τ_i's is not available. Therefore, we cannot use the control variate technique for these parameters, nor evaluate the region of the marginal densities explored by the chains of the τ_i's.

FIGURE 8.7.   Convergence control by the Riemann sum diagnostic for a simulated sample of size 250 for $k = 2$. *Left:* Convergence of the different Rao-Blackwellized Riemann control variates for the different parameters. *Right:* Convergence of the different Rao-Blackwellized Riemann sum estimates for the different parameters. The superimposed graphs correspond to the different choices of ℓ in (8.9). The top graphs correspond to λ, τ_1 and τ_2 and the bottom graphs to q_0 and q_1. (*Source:* Gruet *et al.*, 1998.)

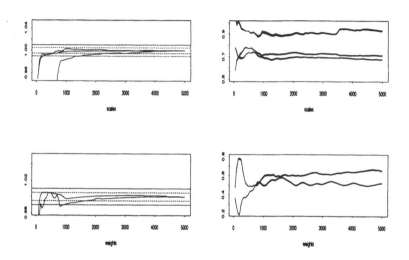

FIGURE 8.8.   Convergence control by Riemann control variate and Rao-Blackwellized Riemann sum estimates for the hospital stay dataset (same legend as Figure 8.7.) (*Source:* Gruet *et al.*, 1998.)

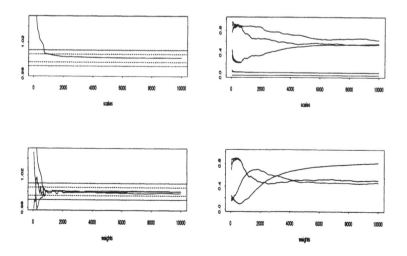

References

Aitkin, M. (1997) Discussion of "On Bayesian Analysis of Mixtures with an unknown Number of Components" by S. Richardson and P. Green. *J. Royal Statist. Soc.* (Ser. B) **59**, 764–766.

Archer, G.E.B. and Titterington, D.M. (1995) Parameter estimation for hidden Markov chains. Tech. report, Dept. of Stat., U. of Glasgow.

Asmussen, S. (1979) *Applied Probability and Queues*. J. Wiley, New York.

Athreya, K.B., Doss, H. and Sethuraman, J. (1996) On the convergence of the Markov chain simulation method. *Ann. Statis.* **24**, 69–100.

Basford, K.E., McLachlan, G.J. and York, M.G. (1998) Modelling the distribution of stamp paper thickness via finite normal mixtures: the 1872 Hidalgo stamp issue of Mexico revisited. *Applied Stat.* (Ser. C) (to appear).

Battacharya, R.N. and Waymire, E.C. (1990) *Stochastic Processes with Applications*. J. Wiley, New York.

Baum, L.E., Petrie, T., Soules, G. and Weiss, N.A. (1970) A maximization technique occuring in the statistical analysis of probabilistic functions of Markov chains. *Ann. Math. Statist.* **41**, 164-171.

Besag, J. (1974) Spatial interaction and the statistical analysis of lattice systems (with discussion). *J. Royal Statist. Soc.* (Ser. B) **36**, 192–326.

Besag, J. (1989) Towards Bayesian image analysis. *J. Applied Statistics* **16**, 395–407.

Besag, J. (1994) Discussion of "Markov chains for exploring posterior distributions". *Ann. Statist.* **22**, 1734-1741.

Besag, J. and Green, P.J. (1992) Spatial Statistics and Bayesian computation (with discussion). *J. Royal Statist. Soc.* (Ser. B) **55**, 25–38.

Besag, J., Green, P.J., Higdon, D. and Mengersen, K.L. (1995) Bayesian computation and stochastic systems (with discussion). *Statis. Science* **10**, 3–66.

Best, N.G., Cowles, M.K. and Vines, K. (1995) CODA: Convergence Diagnosis and Output Analysis software for Gibbs sampling output, Version 0.30. Tech. report, MRC Biostatistics Unit, Univ. of Cambridge.

Billingsley, P. (1968) *Convergence of Probability Measures*. J. Wiley, New York.

Billingsley, P. (1986) *Probability and Measure* (2nd edition). J. Wiley, New York.

Bolthausen, E. (1982) The Berry-Esséen Theorem for Strongly Mixing Harris Recurrent Markov Chains. *Z. Wahrsch. verw. Gebiete* **60**, 283–289.

Bradley, R.C. (1986) Basic properties of strong mixing conditions. In *Dependence in Probability and Statistics*, E. Ebberlein and M. Taqqu (Eds.), 165–192. Birkhäuser, Boston.

Brooks, S.P. (1998) Markov Chain Monte Carlo Method and its Application. *The Statistician* **47**, 69–100.

Brooks, S.P., Dellaportas, P. and Roberts, G.O. (1997) A total variation method for diagnosing convergence of MCMC algorithms. *J. Comput. Graph. Statist.* **6**, 251–265.

Brooks, S.P and Gelman, A. (1998) Alternative methods for monitoring convergence of iterative simulations. *J. Comput. Graph. Statist.* (to appear).

Brooks, S.P. and Roberts, G.O. (1997) On Quantile Estimation and MCMC Convergence. Tech. report, University of Bristol.

Brooks, S.P and Roberts, G. (1998) Diagnosing convergence of Markov chain Monte Carlo algorithms. *Statistics and Computing* (to appear).

Capéraà, P. and van Cutsem, B. (1988) *Méthodes et Modèles en Statistique non Paramétrique*. Dunod, Paris.

Casella, G. and George, E.I. (1992) An introduction to Gibbs sampling. *Amer. Statist.* **46**, 167–174.

Casella, G. and Robert, C.P. (1996) Rao-Blackwellisation of sampling schemes. *Biometrika* **83**(1), 81–94.

Castledine, B. (1981) A Bayesian analysis of multiple-recapture sampling for a closed population. *Biometrika* **67**, 197–210.

Celeux, G. and Clairambault, J. (1992) Estimation de chaînes de Markov cachées: méthodes et problèmes. In *Approches Markoviennes en Signal et Images*, GDR CNRS Traitement du Signal et Images, 5–19.

Celeux, G. and Diebolt, J. (1985) The SEM algorithm: a probabilistic teacher algorithm derived from the EM algorithm for the mixture problem. *Comput. Statist. Quater.* **2**, 73–82.

Chan, K.S. and Geyer, C.J. (1994) Discussion of "Markov chains for exploring posterior distribution". *Ann. Statis.* **22**, 1747–1758.

Chauveau, D. and Diebolt, J. (1997) MCMC convergence diagnostic via the Central Limit Theorem. Preprint # 22/97, Université Marne-la-Vallée.

Chen, M.H. and Shao, Q.M. (1997) On Monte Carlo Methods for Estimating Ratios of Normalizing Constants. *Ann. Statis.* **25**, 1563–1594.

Chib, S. and Greenberg, E. (1995) Understanding the Metropolis-Hastings algorithm. *Amer. Statist.* **49**, 327–335.

Chung, K.L. (1967) *Markov Processes with Stationary Transition Probabilities*. Springer-Verlag, Heidelberg.

Churchill, G.A. (1989) Stochastic models for heterogeneous DNA sequences. *Bull. Math. Biol.* **51**, 79–94.

Cowles, M.K. and Carlin, B.P. (1996) Markov Chain Monte-Carlo convergence diagnostics: a comparative study. *J. Amer. Statist. Assoc.* **91**, 883–904.

Dacunha-Castelle, D. and Duflo, M. (1986) *Probability and Statistics*, vol.

II. Springer-Verlag, New York.

Damien, P. and Walker, S. (1996) Sampling probability densities via uniform random variables and a Gibbs sampler. Preprint.

Davydov, Y.A. (1973) Mixing conditions for Markov chains. *Theory Probab. Appl.* **18**, 312–328.

Dempster, A.P., Laird, N.M. and Rubin, D.B. (1977) Maximum likelihood from incomplete data via the EM algorithm (with discussion). *J. Royal Statist. Soc.* (Ser. B) **39**, 1–38.

Denison, D.G.T., Mallick, B.K. and Smith, A.F.M. (1998) Automatic Bayesian curve fitting. *J. Royal Statist. Soc.* (Ser. B) **60**, 333–350.

Devroye, L. (1985) *Non-Uniform Random Variate Generation*. Springer-Verlag, New York.

Diebolt, J. and Robert, C.P. (1990) Estimation des paramètres d'un mélange par échantillonnage bayésien. *Notes aux Comptes-Rendus de l'Académie des Sciences I* **311**, 653–658.

Diebolt, J. and Robert, C.P. (1993) Discussion of "Bayesian computations via the Gibbs sampler" by A.F.M. Smith and G. Roberts. *J. Royal Statist. Soc.* (Ser. B) **55**, 71–72.

Diebolt, J. and Robert, C.P. (1994) Estimation of finite mixture distributions by Bayesian sampling. *J. Royal Statist. Soc.* (Ser. B) **56**, 363–375.

Dupuis, J.A. (1995) Bayesian estimation of movement probabilities in open populations using hidden Markov chains. *Biometrika* **82**, 761–772.

Feller, W. (1970) *An Introduction to Probability Theory and its Applications.*, Vol. 1. J. Wiley, New York.

Feller, W. (1971) *An Introduction to Probability Theory and its Applications.*, Vol. 2. J. Wiley, New York.

Fill, J.A. (1998a) An Interruptible Algorithm for Exact Sampling via Markov Chains. *Ann. Applied Prob.* (to appear).

Fill, J.A. (1998b) The Move-to Front Rule: A Case study for Two Perfect Sampling Algorithms. *Proba. Eng. Info. Scis.* (to appear).

Freydman, H. (1992) A non parametric estimation procedure for a periodically observed three-state Markov process, with application to AIDS. *J. Royal Statist. Soc.* (Ser. B) **54**, 853–866.

Gamerman, D. (1997) *Markov Chain Monte Carlo*. Chapman and Hall, London.

Gaver, D.P. and O'Muircheartaigh, I.G. (1987) Robust empirical Bayes analysis of event rates. *Technom.* **29**, 1–15.

Gelfand, A.E. and Smith, A.F.M. (1990) Sampling based approaches to calculating marginal densities. *J. Amer. Statist. Assoc.* **85**, 398–409.

Gelfand, A.E. and Smith, A.F.M. (1998) *Bayesian Computation*. J. Wiley (to appear).

Gelman, A., Gilks, W.R. and Roberts, G.O. (1996) Efficient Metropolis jumping rules. In *Bayesian Statistics 5*, J.O. Berger, J.M. Bernardo, A.P. Dawid, D.V. Lindley and A.F.M. Smith (Eds.). Oxford University Press, Oxford, 599–608.

Gelman, A. and Rubin, D.B. (1992) Inference from iterative simulation using multiple sequences (with discussion). *Statis. Science* **7**, 457–511.

Geman, S. and Geman, D. (1984) Stochastic relaxation, Gibbs distributions and the Bayesian restoration of images. *IEEE Trans. Pattern Anal. Mach. Intell.* **6**, 721–741.

Gentleman, R. C., Lawless, J. F., Lindsey, J. C. and Yan, P. (1994) Multi-stage Markov models for analysing incomplete disease history data with illustrations for HIV Disease. *Statist. Med.* **13**, 805–821.

George, E.I. and Robert, C.P. (1992) Calculating Bayes estimates for capture-recapture models. *Biometrika* **79**(4), 677–683

Geweke, J. (1992) Evaluating the accuracy of sampling-based approaches to the calculation of posterior moments (with discussion). In *Bayesian Statistics 4*, J.M. Bernardo, J.O. Berger, A.P. Dawid and A.F.M. Smith (Eds.), 169–193. Oxford University Press, Oxford.

Geyer, C.J. (1992) Practical Monte Carlo Markov Chain (with discussion). *Statis. Science* **7**, 473–511.

Geyer, C.J. (1995) Conditioning in Markov Chain Monte Carlo. *J. Comput. Graph. Statis.* **4**, 148–154.

Ghosh J. (1996) A new graphical tool to detect non normality. *J. Royal Statist. Soc.* (Ser. B) **58**, 691–702.

Gilks, W.R., Best, N.G. and Tan, K.K.C. (1995) Adaptive Rejection Metropolis Sampling within Gibbs Sampling. *Applied Statist.* (Ser. C) **44**, 455–472.

Gilks, W.R., Richardson, S. and Spiegelhalter, D.I. (1996) *Markov Chain Monte Carlo in Practice*. Chapman and Hall, London.

Gilks, W.R. and Roberts, G.O. (1996) Strategies for improving MCMC. In *Markov Chain Monte Carlo in Practice*, W.R. Gilks, S. Richardson, and D.J. Spiegelhalter (Eds.), 89–114. Chapman and Hall, London.

Gilks, W.R., Roberts, G.O. and Sahu, S.K. (1998) Adaptive Markov Chain Monte Carlo. *J. Amer. Statist. Assoc.* (to appear).

Green, P.J. (1995) Reversible jump MCMC computation and Bayesian model determination. *Biometrika* **82**(4), 711–732.

Green, P.J. and Murdoch, D. (1998) Exact sampling for Bayesian inference: towards general purpose algorithms. In *Bayesian Statistics 6*, J.O. Berger, J.M. Bernardo, A.P. Dawid, D.V. Lindley and A.F.M. Smith (Eds.). Oxford University Press, Oxford (to appear).

Grenander, U. and Miller, M. (1994) Representations of Knowledge in Complex Systems (with discussion). *J. Royal Statist. Soc.* (Ser. B) **56**, 549–603.

Gruet, M.A., Philippe, A. and Robert, C.P. (1998) MCMC control spreadsheets for exponential mixture estimation. Doc. travail DT9808, CREST, Insee, Paris.

Guihenneuc–Jouyaux, C. and Richardson, S. (1998) Modèle de Markov avec erreurs de mesure : approche bayésienne et application au suivi longitudinal des lymphocytes T4. In *Biométrie et Applications Bayésiennes*, Economica, Paris (to appear).

Guihenneuc–Jouyaux, C. and Robert, C.P. (1998) Finite Markov chain convergence results and MCMC convergence assessment. *J. Amer. Statist. Assoc.* (to appear).

Hammersley, J.M. (1974) Discussion of Besag's paper. *J. Royal Statist. Soc.* (Ser. B) **36**, 230–231.

Hastings, W.K. (1970) Monte Carlo sampling methods using Markov chains and their application. *Biometrika* **57**, 97–109.

Heidelberger, P. and Welch, P.D. (1983) A spectral method for confidence interval generation and run-length control in simulation. *Commun. ACM* **24**, 233–245.

Heitjan, D.F. and Rubin, D.B. (1991) Ignorability and coarse data. *Ann. Statis.* **19**, 2244–2253.

Hills, S.E. and Smith, A.F.M. (1992) Parametrization issues in Bayesian inference. In *Bayesian Statistics 4*, Ed. J.M. Bernardo, J.O. Berger, A.P. Dawid and A.F.M. Smith. Oxford University Press, Oxford, 641-649.

Hills, S.E. and Smith, A.F.M. (1993) Diagnostic plots for improved parameterization in Bayesian inference. *Biometrika* **80**, 61–74.

Hobert, J.P., Robert, C.P. and Goutis, C. (1996) Connectedness conditions for the convergence of the Gibbs sampler. *Statis. Prob. Letters* **33**, 235–240.

Izenman, A.J. and Sommer, C.J. (1988) Philatelic mixtures and multimodal densities. *J. Amer. Statist. Assoc.* **83**, 941–953.

Jensen, S.T., Johansen, S. and Lauritzen, S.L. (1991) Globally convergent algorithms for maximizing a likelihood function. *Biometrika* **78**(4), 867-877.

Johnson V.E. (1996) Studying convergence of Markov chain Monte Carlo algorithms using coupled sample paths. *J. Amer. Statist. Assoc.* **91**, 154–166.

Juang, B.H. and Rabiner, L.R. (1991) Hidden Markov models for speech recognition. *Technom.* **33**, 251–272.

Kemeny, J.G. and Snell, J.L. (1960) *Finite Markov Chains.* Springer-Verlag, New York.

Kendall, W. (1998) Perfect Simulation for the Area-Interaction Point Process. In *Probability Perspective*, Heyde, C.C. and Accardi, L. (Eds.). World Scientific Press (to appear).

Kirby, A. J. and Spiegelhalter, D. J. (1994) Statistical Modeling for the Precursors of Cervical Cancer. In *Case Studies in Biometry*, N. Lange (Ed.). John Wiley, New York.

Lehmann, E.L. and Casella, G. (1998) *Theory of Point Estimation* (revised edition). Chapman and Hall, New York (to appear).

Lezaud, P. (1998) Chernoff bound for finite Markov chains. *Ann. Applied Proba.* (to appear).

Lindvall, T. (1992) *Lectures on the Coupling Theory.* J. Wiley, New York.

Liu, J.S. (1995) Metropolized Gibbs Sampler: An improvement. Tech. report, Dept. of Statistics, Stanford U., California.

Liu, C., Liu, J.S., and Rubin, D.B. (1992) A variational control for assessing the convergence of the Gibbs sampler. In *Proceedings of the American*

Statistical Association, Statistical Computing Section, 74–78.

Liu, J.S., Wong, W.H. and Kong, A. (1994) Covariance structure of the Gibbs sampler with applications to the comparisons of estimators and sampling schemes. *Biometrika* **81**, 27–40.

Liu, J.S., Wong, W.H. and Kong, A. (1995) Correlation structure and convergence rate of the Gibbs sampler with various scans. *J. Royal Statist. Soc.* (Ser. B) **57**, 157–169.

Longini, I., Clark, W. S., Gardner, L. I. and Brundage, J. F. (1991) The Dynamics of CD4+ T-lymphocyte decline in HIV-infected individuals: A Markov modeling approach. *J. AIDS* **4**, 1141–1147.

Longini, I., Clark, W. S. and Karon, J. (1993) The effect of routine use of therapy on the clinical course of HIV infection in a population-based cohort. *Am. J. Epidemiol.* **137**, 1229–1240.

Mann, B. (1996) *Berry-Esséen Central Limit Theorem for Markov Chains.* PhD Thesis, Harvard University.

McKeague, I.W. and Wefelmeyer, W. (1995) Markov Chain Monte Carlo and Rao-Blackwellization. Tech. report, Florida State University.

Mengersen, K.L. and Robert, C.P. (1996) Testing for mixtures: a Bayesian entropic approach (with discussion). In *Bayesian Statistics 5*, J.O. Berger, J.M. Bernardo, A.P. Dawid, D.V. Lindley and A.F.M. Smith (Eds.). Oxford University Press, Oxford, 255–276.

Mengersen, K.L., Robert, C.P. and Guihenneuc-Jouyaux, C. (1998) MCMC convergence diagnostics: a "reviewwww". In *Bayesian Statistics 6*, J.O. Berger, J.M. Bernardo, A.P. Dawid, D.V. Lindley and A.F.M. Smith (Eds.). Oxford University Press, Oxford (to appear).

Mengersen, K.L. and Tweedie, R.L. (1996) Rates of convergence of the Hastings and Metropolis algorithms. *Ann. Statist.* **24**, 101–121.

Metropolis, N., Rosenbluth, A.W., Rosenbluth, M.N., Teller, A.H. and Teller, E. (1953) Equations of state calculations by fast computing machines. *J. Chem. Phys.* **21**, 1087–1092.

Meyn, S.P. and Tweedie, R.L. (1993) *Markov Chains and Stochastic Stability.* Springer-Verlag, London.

Mira, A. and Geyer, C.J. (1998) Ordering Monte Carlo Markov chains. Tech. report, U. Minnesota.

Müller, P. (1991) A generic approach to posterior integration and Gibbs sampling. Tech. Report # 91-09, Purdue Uni., West Lafayette, Indiana.

Müller, P. (1993) Alternatives to the Gibbs sampling scheme. Tech. report, Institute of Statistics and Decision Sciences, Duke Uni.

Murdoch, D.J. and Green, P.J. (1998) Exact Sampling for a Continuous State. *Scandinavian J. Statist.* (to appear).

Muri, F. (1997) *Comparaison d'algorithmes d'identification de chaînes de Markov cachées et application à la détection de régions homogènes dans les séquences d'ADN.* PhD thesis, Université René Descartes, Paris.

Muri, F. (1998) Modelling Bacterial Genomes using Hidden Markov Models. In *Compstat' 98*, F. Payne (Ed.). Physica-Verlag, Berlin (to appear).

Mykland, P., Tierney, L. and Yu, B. (1995) Regeneration in Markov chain samplers. *J. Amer. Statist. Assoc.* **90**, 233–241.

Neal, R.M. (1993) *Probabilistic Inference using Markov Chain Monte Carlo Methods*. Dept. of Computer Science, U. of Toronto.

Orey, S. (1971) *Limit Theorems for Markov Chain Transition Probabilities*. Van Nostrand, London.

Peskun, P.H. (1973) Optimum Monte-Carlo sampling using Markov chains. *Biometrika* **60**, 607–612.

Philippe, A. (1997a) Processing simulation output by Riemann sums. *J. Statist. Comput. Simulation* **59**, 295–314.

Philippe, A. (1997b) Importance sampling and Riemann Sums. *Prepub. IRMA Lille* **43**,VI.

Philippe, A. (1997c) Simulation of right and left truncated Gamma distributions by mixtures. *Statistics and Computing* **7**, 173-181.

Philippe, A. and Robert, C.P. (1998) Riemann sums for MCMC estimation and control. Tech. report, Univ. Rouen.

Phillips, D.B. and Smith, A.F.M. (1996) Bayesian model comparison via jump diffusions. In *Markov Chain Monte Carlo in Practice*, W.R. Gilks, S. Richardson, and D.J. Spiegelhalter (Eds.), 215–240. Chapman and Hall, London.

Propp, J.G. and Wilson, D.B. (1996) Exact sampling with coupled Markov chains and applications to statistical mechanics. *Random Structures and Algorithms* **9**, 223-252.

Qian, W. and Titterington, D.M. (1990) Parameter estimation for hidden Gibbs chains. *Statis. Prob. Letters* **10**, 49-58.

Qian, W. and Titterington, D.M. (1991) Estimation of parameters in hidden Markov models. *Phil. Trans. Roy. Soc. London* A **337**, 407–428.

Qian, W. and Titterington, D.M. (1992) Stochastic relaxations and EM algorithms for Markov random fields. *J. Statist. Comput. Simulation* **40**, 55–69.

Rabiner, L.R. (1989) Tutorial on Hidden Markov Models and Selected Applications in Speech Recognition. *Proceedings of the IEEE* **77**, 257-286.

Raftery, A.E. and Lewis, S. (1992a) How many iterations in the Gibbs sampler? In *Bayesian Statistics 4*, J.O. Berger, J.M. Bernardo, A.P. Dawid and A.F.M. Smith (Eds.), 763–773. Oxford University Press, Oxford.

Raftery, A.E. and Lewis, S. (1992b) The Number of Iterations, Convergence Diagnostics and Generic Metropolis Algorithms. Tech. report, Department of Statistics, U. of Washington, Seattle.

Raftery, A.E. and Lewis, S. (1996) Implementing MCMC. In *Markov Chain Monte Carlo in Practice*, W.R. Gilks, S. Richardson, and D.J. Spiegelhalter (Eds.), 115–130. Chapman and Hall, London.

Revuz, D. (1984) *Markov Chains* (2nd edition). North-Holland, Amsterdam.

Richardson, S. and Green, P.J. (1997) On Bayesian analysis of mixtures with an unknown number of components (with discussion). *J. Royal Statist.*

Soc. (Ser. B) **59**, 731–792.

Richardson, S. and Guihenneuc-Jouyaux, C. (1996) Contribution to the discussion ot the paper by Satten and Longini. *Applied Statistics* (Ser. C) **45**, 298–299.

Ripley, B.D. (1987) *Stochastic Simulation.* J. Wiley, New York.

Robert, C.P. (1993) Prior Feedback: A Bayesian approach to maximum likelihood estimation. *Comput. Statist.* **8**, 279–294.

Robert, C.P. (1994a) *The Bayesian Choice.* Springer-Verlag, New York.

Robert, C.P. (1994b) Discussion of "Markov chains for exploring posterior distribution". *Ann. Statis.* **22**, 1742–1747.

Robert, C.P. (1995) Simulation of truncated normal variables. *Statistics and Computing* **5**, 121–125.

Robert, C.P. (1996a) Convergence control techniques for Markov Chain Monte Carlo algorithms. *Statis. Science* **10**, 231–253.

Robert, C.P. (1996b) Inference in mixture models. In *Markov Chain Monte Carlo in Practice*, W.R. Gilks, S. Richardson, and D.J. Spiegelhalter (Eds.), 441–464. Chapman and Hall, London.

Robert, C.P. (1996c) *Méthodes de Monte Carlo par Chaînes de Markov.* Economica, Paris.

Robert, C.P. (1997) Discussion of Richardson and Green's paper. *J. Royal Statist. Soc.* (Ser. B) **59**, 758–764.

Robert, C.P. and Casella, G. (1998) *Monte-Carlo Statistical Methods.* Springer-Verlag, New York (to appear).

Robert, C.P., Celeux, G. and Diebolt, J. (1993) Bayesian estimation of hidden Markov models: a stochastic implementation. *Statis. Prob. Letters* **16**, 77–83.

Robert, C.P. and Mengersen, K.L. (1998) Reparametrization issues in mixture estimation and their bearings on the Gibbs sampler. *Comput. Statis. Data Ana.* (to appear).

Robert, C.P., Rydén, T. and Titterington, D.M. (1998) Convergence controls for MCMC algorithms, with applications to hidden Markov chains. Tech. report, Uni. of Glasgow.

Robert, C.P. and Titterington, M. (1998) Resampling schemes for hidden Markov models and their application for maximum likelihood estimation. Tech. *Statistics and Computing* (to appear).

Roberts, G.O. (1992) Convergence diagnostics of the Gibbs sampler. In *Bayesian Statistics 4*, J.M. Bernardo, J.O. Berger, A.P. Dawid and A.F.M. Smith (Eds.), 775–782. Oxford University Press, Oxford.

Roberts, G.O. (1994) Methods for estimating L^2 convergence of Markov chain Monte Carlo. In *Bayesian Statistics and Econometrics: Essays in Honor of Arnold Zellner*, D. Barry, K. Chaloner and J. Geweke (Eds.). J. Wiley, New York.

Roberts, G.O., Gelman, A. and Gilks, W.R. (1995) Weak convergence and optimal scaling of random walk Metropolis algorithms. Res. report 94–16, Stat. Lab., U. Cambridge.

Roberts, G.O. and Rosenthal, J.S. (1997) Markov chain Monte Carlo: some practical implications of theoretical results. Tech. report, Stats. Lab., U. of Cambridge.

Roberts, G.O. and Tweedie, R.L. (1996) Geometric convergence and Central Limit Theorems for multidimensional Hastings and Metropolis algorithms. *Biometrika* **83**, 95–110.

Rosenblatt, M. (1971) *Markov Processes: Structure and Asymptotic Behavior.* Springer-Verlag, New York.

Saloff-Coste, L. and Diaconis, P. (1993) Comparison theorems for reversible Markov chains. *Ann. Appl. Prob.* **3**, 696–730.

Satten, G. A. and Longini, I. M. (1996) Markov Chains with Measurement Error : Estimating the True Course of a marker of the Progression of Human Immunodeficiency Virus Disease. *Appl. Stat.* (Ser. C) **45**, 275–309.

Schruben, L., Singh, H. and Tierney, L. (1983) Optimal tests for initialization bias in simulation output. *Operation. Research* **31**, 1176–1178.

Seoh, M. and Hallin, M. (1997) When does Edgeworth beat Berry and Esséen? *J. Statist. Plann. Infer.* (to appear).

Shapiro, S.S. and Wilk, M.B. (1965) An analysis of variance test for normality. *Biometrika* **52**, 591–611.

Smith, A.F.M. and Roberts, G.O. (1993) Bayesian computation via the Gibbs sampler and related Markov chain Monte Carlo methods (with discussion). *J. Royal Statist. Soc.* (Ser. B) **55**, 3–24.

Tanner, M. (1996) *Tools for Statistical Inference: Observed Data and Data Augmentation Methods* (3rd edition). Springer-Verlag, New York.

Tanner, M. and Wong, W. (1987) The calculation of posterior distributions by data augmentation. *J. Amer. Statist. Assoc.* **82**, 528–550.

Tierney, L. (1994) Markov chains for exploring posterior distributions (with discussion). *Ann. Statist.* **22**, 1701–1786.

Titterington, D.M., Smith, A.F.M. and Makov, U.E. (1985) *Statistical Analysis of Finite Mixture Distributions.* J. Wiley, New York.

Wolter, W. (1986) Some coverage error models for census data. *J. Amer. Statist. Assoc.* **81**, 338–346.

Yakowitz, S., Krimmel, J.E. and Szidarovszky, F. (1978) Weighted Monte-Carlo integration. *SIAM J. Numer. Anal.* **15**(6), 1289–1300.

Yu, B. (1995) Discussion to Besag *et al.* (1995). *Statist. Sci.* **10**, 54–58.

Yu, B. and Mykland, P. (1998) Looking at Markov Samplers Through Cusum Path Plots: A Simple Diagnostic Idea. *Statistics and Computing* (to appear).

Author Index

Subject Index

Lecture Notes in Statistics

For information about Volumes 1 to 61
please contact Springer-Verlag